Identified skeletal collections: the testing ground of anthropology?

Edited by

Charlotte Yvette Henderson
Francisca Alves Cardoso

Archaeopress Publishing Ltd
Summertown Pavilion
18-24 Middle Way
Summertown
Oxford OX2 7LG

www.archaeopress.com

ISBN 978 1 78491 805 7
ISBN 978 1 78491 806 4 (e-Pdf)

Printed in England by Oxuniprint, Oxford

This book is available direct from Archaeopress or from our website www.archaeopress.com

Contents

Chapter 6. The significance of identified human skeletal collections to further our understanding of the skeletal ageing process in adults 115

Vanessa Campanacho and Hugo F.V. Cardoso

Chapter 7. Secular changes in cranial size and sexual dimorphism of cranial size: a comparative analysis of standard cranial dimensions in two Portuguese identified skeletal reference collections and implications for sex estimation.. 133

Luísa Marinho, Ana R. Vassalo and Hugo F. V. Cardoso

Chapter 1

Introduction

Charlotte Henderson

CIAS – Research Centre for Anthropology and Health, Department of Life Sciences, University of Coimbra, Portugal.

This chapter aims to give a broad overview of the topics covered in this book, along with a background and discussion of what identified skeletal collections are. There are two themes which run through the book: uses of identified collections and ethical issues surrounding their existence and use. These themes are broad and the chapters in the book are therefore wide-ranging, covering everything from descriptions of specific collections (Chapters 2 and 3), to relating data from living people with that associated with collections (Chapters 3 and 8), via the testing of ageing methods (Chapters 5, 6 and 7), to problems of biased collections (Chapters 4, 5 and 9). They also cover the collection of the skeletons within both a legal framework and as creations with an anthropological purpose. It is important to remember throughout this introduction and the rest of the book that these skeletons represent once living people and that these remains and the biographical data associated with them (including age, sex or gender and cause of death) need to be treated with respect and dignity. In providing an overview of this book, this chapter will also discuss what respect and dignity may mean in this context and how research, dissemination (academic and to the general public) and new technologies (including social media) may impact on this. The overall objective of this chapter is to make the reader think critically about these issues.

What are identified human skeletal collections?

It will be seen that each chapter contains a slightly different definition or description of what these collections are. This is important, but it strikes at one of the main issues: if there is no definition of what they are then there can be no good discussion of their existence, use or treatment. For the purposes of this book, the aim was to keep the definition broad to be as inclusive as possible. This book covers identified skeletal collections derived from archaeological sources (Chapter 2 and 5), from dissections and other cadaver sources (Chapters 3 and 9) and those created from recent (or currently used) cemetery sites specifically for anthropological (or bioarchaeological or osteological) research (Chapters 7 and 8). This disparate starting point means that identified skeletons can be found in a wide range of differing environments (museums, university departments and churches) and be covered by different legislation (aside from country specific differences which also exist) as well as different codes of both practice and ethics (as will be discussed by the authors in this book). Almost universally (in the case of these collections), these skeletons are treated as anonymous individuals who have specific (and useful) data associated with them: normally age and sex (although it may be important to consider whether the sex of the individual has been socially constructed). However, it should be borne in mind that some skeletons are not discussed as anonymous individuals and this is predominantly the case for 'famous' or historically 'important' individuals, e.g. King Richard III of England, Henry VII of Luxembourg, the castrato singer Farinelli, and the novelist Samuel Richardson (Belcastro et al., 2011; Buckley et al., 2013; Mariotti et al., 2013; Scheuer et al., 1994; Scorrano et al., 2017).

The source of the skeletons causes biases in the structure of the collections (see specifically Chapters 4 and 5, but discussed throughout the book). The archaeologically derived collections are are often of those who could afford burial in such a way that they could be identified several hundred years later, e.g. they are buried with coffin plates or in a crypt (Caffell & Holst, 2010; Cox, 1995; Molleson & Cox, 1993; Scheuer & Bowman, 1995). In contrast, those derived from recent cemeteries are often considered to be of lower socio-economic status backgrounds (for details and a more in depth discussion of this see Chapters 6 and 7). Whereas those derived from anatomical dissections come from either end of the socio-economic scale, depending on the conditions during the period when the collection was amassed (Chapter 4). Furthermore, it can impact on the availability of documentary data and the information available (and quality of it) concerning each individual in the collection. Typically, as will be seen in this book, data tend not to be restricted to age and sex. Other information, such as name, address, occupation, place of death, date of death, cause of death, marital status and birth date are also included. This ensures that collections of these individuals are ideal for testing a variety of hypotheses and developing new recording methods (see below). However, limitations in the availability of the data as well as their quality and reliability does have an impact on the use of the collections (see Chapters 4, 5, and 8). The size of these collections is also variable, again in part depending on the source or purpose of the original collection. Much of the biographical data collected and how they were collected is based on contemporaneous anthropological views and for this reason 'race' or ethnicity is also listed, particularly in North America and South Africa (see Chapters 4, 5 and most specifically 9).

While this book focusses on a limited geographical area: Britain, Canada, Portugal, South Africa and the USA. It should be noted that other countries hold similar collections derived from the same type of sources. Therefore, it is important to remember that the collections and the consideration surrounding them are not limited to a few countries or continents (Alemán et al., 2012; Belcastro et al., 2017; Chi Keb et al., 2017; Eliopoulos et al., 2007; Facchini et al., 2006; Rissech & Steadman, 2011; Ross & Manneschi, 2011; Salceda et al., 2012; Ubelaker, 2014). This is a further bias both within this book and in numerous other publications: a fixation on a limited range of collections used mostly because they are well known, easy to access, and (aside from the Portuguese collections) are mostly in English speaking countries. Beyond the geographical limitations there are also temporal aspects: most (although not all) collections consist of individuals who died during the past 150 years. Therefore they only provide a very short temporal snapshot of the human skeleton. The British identified collections (Chapters 2 and 5) include individuals from earlier times but they too are limited to the post-mediaeval period. These biases will be discussed below.

Conception of the book

This book was originally conceived after a conference session '*Identified Skeletal Collections: The testing ground of anthropology?*' presented at the 17th World Congress of the International Union of Anthropological and Ethnological Sciences in Manchester, in August 2013. The session was organised due to the number of studies being undertaken on identified skeletal collections and a lack of discussion surrounding their existence and use. These collections, as will become very clear in the subsequent chapters, are very widely used (particularly by students) and there is often reanalysis of the same collections to test the same (or the same type of) questions and this is reflected within this book. Reanalysis leads to repeated handling of remains and it is well known that re-handling can lead to deterioration of the remains (Caffell et al., 2001). Collection management is, therefore, also an important consideration. It is also important to remember that, as the field develops, new technical methods develop, e.g. three-dimensional scanning and printing as well as studies of DNA. This means new questions can be asked, while new methods also need to be tested and compared to older ones meaning that re-analysis can be necessary. In addition to this it is also necessary to consider how this research is disseminated to the broader public, for example through museum displays, academic and non-academic publications and, increasingly digitally via websites, blogs and social media.

Within the context of this book, the main aspect to be discussed was the use of identified skeletal collections in relation to the ethical aspects involved in their curation (and possibly original provenance). These collections straddle a limbo between the skeleton (rarely identified to a particular individual) and the fleshed cadaver. Unlike fleshed cadavers (IFAA, 2011), which are also identifiable, there has been little discussion about how research on these skeletons should be undertaken and what should be reported. There is therefore a need for discussion whether such guidelines are needed, what they should cover and how often they should be updated. Whether these guidelines need to be different to those covering non-identified remains also needs to be considered or whether the same guidelines (perhaps derived from guidelines and codes of practice for anatomical cadavers) can be applied to all remains. If guidelines are needed, then should they be local to a country, a type of collection, or global basic standard? As can be seen from the chapters, local conditions affect what is already in place. It should be noted at this point that this is a different question to whether new legislation is needed to cover these collections, which (as can be seen in this book but see also (Márquez-Grant & Fibiger, 2011) are already covered by existing laws and precedents.

Why are they so useful?

As stated above identified collections are very widely used, particularly by students and the first step in considering ethics is to understand why they are so useful. Every chapter of this book discusses the uses and importance of these collections for anthropological and archaeological research. Numerous specific and general examples of these types of research are covered throughout the book, so this introduction will only consider these broadly.

Starting with the forensic importance of these collections, it may be thought that DNA analysis has superseded morphometric skeletal analysis for identifying skeletal human remains. However, DNA does not survive in all burial environments (Collins et al., 2002). Comparative DNA is also needed for identification purposes which may not always be available. For these reasons morphometric methods to estimate both age and sex are needed, and they need to be appropriate for the population (Ross & Manneschi, 2011). This need for population specific approaches is discussed below, as well as in detail in Chapter 4. Methods for estimating age and sex are also needed in archaeological contexts. Without such methods (as well as other osteological approaches including palaeopathological) it would, for example, have been an expensive guess to do confirmatory DNA testing on the skeleton identified as Richard III (Buckley et al., 2013). In more normal archaeological contexts without such methods it would be impossible to undertake any demographic research or to study the effect of disease on population health: particularly which groups were more vulnerable to specific diseases.

These collections also play an important role in interpreting skeletal changes and for identifying diseases in skeletal remains (some examples of this include, Cardoso, 2008; Henderson et al., 2013; Santos & Roberts, 2006). Clinical data can be used as a foundation, but this rarely focusses on the bone changes and some diseases no longer progress to causing bone changes due to modern interventions, e.g. antibiotics. Therefore studying pre-antibiotic era skeletons and focussing on hard tissue changes provides details not readily available in clinical literature. This, for example, has been a particular focus of my own research studying the impact of occupations on muscle attachment sites to the skeleton: attachment sites which are primarily studied in relation to rheumatological conditions in modern clinical literature (Henderson et al., 2017). Such aspects are particularly important for archaeological research aiming to reconstruct lives in the past, including health, disease and disability. Here the other biographical data associated with the skeletons becomes more important. Cause of death can assist in differentiating lesions occurring in similar regions but with different causative agents, e.g. pulmonary tuberculosis and bronchitis (Santos & Roberts, 2006). Data on occupation can also enable morphological changes, such as those at joints or muscle attachments, to be studied in relation to the working life (Alves-Cardoso & Henderson, 2013). These type of data have more limiting factors than those of age and sex, as they are deeply grounded in governmental aims and broader socio-cultural perceptions as well as scientific knowledge. Two aspects of these limitations are focussed on in this book: namely the designation of 'race' (Chapter 9, but other chapters also deal with this) and occupation (Chapter 8).

Biases

The limitations outlined above are only part of a broader set of biases which occur in these collections. These biases are important to consider not only in terms of why they occur, but the impact on the data collected. This is not the first discussion of biases present in skeletal collections (see Herring & Saunders, 1995), but these remain important and sometimes overlooked issues. Biases are discussed in all chapters, but in this section I will draw your attention to certain types of bias.

Both temporal and geographic biases have been alluded to above. The biggest bias in this book, as well as within each collection, is that of time. The majority of collections discussed in this book were built up in the twentieth century, typically of individuals alive in the early part of that century. This is a general trend for most of the collections which exist. Britain is an unusual exception in this respect with most of the identified collections coming from cemeteries which have long since ceased to be used for burial, e.g. the collection from St. Bride's Church in Fleet Street and the more well known Spitalfields collection (see Chapters 2 and 5 as well as Cox, 1995; Scheuer & Bowman, 1995). Such skeletons may be ideal for testing methods applicable to the pre-modern world. Changes in stature, demographics and medical practices, however, mean that these collections may be less appropriate for the development and testing of forensic standards. Chapter 7 discusses the needs of modern forensics particularly the need for standards to be tested on more recent identified collections: more recent even than those of the early- to mid-twentieth century.

There is often a belief that anthropological methods need to be developed that are appropriate to specific populations, a specific example of this is stature methods for which exist for many different populations (examples of which include Cardoso & Gomes, 2009; Mays, 2016; Ross & Manneschi, 2011). In the past the concept of 'race' was an important reason for this. As discussed in many of the chapters, this concept also shaped the structure and demographic profile of some of these collections. Skeletal differences were often attributed to 'race' in the past, while the terms ethnicity and ancestry are often used today. In several chapters (but specifically 4, 5 and 9) the importance of differences in socio-economic status leading to differential access to resources is considered with relation to these skeletal differences. These socio-economic differences form another bias both within and between collections. Not only do these differences have to be disentangled from geographic, but also temporal variation. It is also important to note that identifying an individual's (or a population's) socio-economic status or their access to resources based solely on the documentary data held within an identified collection is limited by the historical evidence (Alves Cardoso et al., 2016). Life-course biographical data, i.e. data describing an individual's whole life, are typically very limited, as discussed in Chapter 8. However, geographical biases should not be entirely discounted, but these also need to be considered in relation to environmental differences, particularly local disease-burdens.

The final obvious bias is demographic: these collections (as with most cemetery or archaeologically-derived skeletal assemblages) rarely represent a normal living population (Walker, 1995). Furthermore, the documentary data, particularly age, can be biased by how it is collected. Socio-cultural biases can play a role here, particularly in how the different sexes are viewed with females often under-represented in collections created from dissection rooms (see Chapters 3 – 5). A comment also needs to be made about the difference between biological sex and gender. Expressed very simplistically, gender is how we ourselves or society perceive us, whereas biological sex is genetically determined. It is therefore important to consider how the collection was created and where the documentary data come from when using the term 'sex' and when testing sex estimation methods. Similarly the source of documentary data may affect the age ascribed to some skeletons (most chapters discuss this issue). This is particularly important when testing ageing methods or when studying skeletal changes associated with the ageing process.

These, and other biases and limitations, should always be considered when developing and testing methods for analysing skeletal remains. However, these are human remains and not just materials to be used and for this reason it is also important to consider the ethics of their study.

Identified but anonymous

In this section I want to consider the anonymous nature of these collections and how they are treated as collections rather than as individuals. Unlike the skeletons of 'famous' individuals who are discussed (both in the media and in academic publications) by name, these skeletons remain anonymous. Rather than 'refleshing' their lives through skeletal analysis combined with historical documents, as with these 'famous' individuals, their lives are typically considered at a population level. Exceptions to this do exist where osteobiographies of individuals in collections have been published (e.g. Lopes et al., 2010; Scheuer et al., 1994) and where names have been (I think in hindsight, inappropriately) published (Henderson et al., 2013). The International Federations of Anatomist guidelines state that donors should normally remain anonymous (IFAA, 2011) and this is typical for most medical publishing, although there is an acknowledgement that this can be hard to achieve when images or unusual conditions are published (BMJ, n.d.). Should this guideline also be in place for individuals represented in these collections, whatever their initial provenance? Or is the removal of identity (via anonymisation) a continuation of socio-economic differences and deprivations in death? Does this also misrepresent archaeology to the general public, giving the impression that there is a greater focus on the 'famous' than the general population?

Continuing social stratification from life into death in these collections has previously been discussed in detail (Muller et al., 2017) and also Chapter 9 of this book. Naming individuals in these collections would not change this. However, using the research undertaken to reflesh the skeleton can be a good way to engage the wider community via outreach projects, as discussed in Chapter 2 and 9. As discussed in this latter chapter, this can be a good approach to encouraging currently under-represented groups into academia.

New technologies should also dissuade those keen to publish names of individuals in these collections. Three-dimensional representations already exist on the internet in the public domain, images of bones are shared via social media and three-dimensional printing means that (if the point cloud data are available or can be reconstructed) these could be printed by anybody with the resources available. Guidelines for the new digital era are needed and are currently being developed for the UK for the British Association for Biological Anthropology and Osteoarchaeology. Advances in genetics and biomolecular research, alongside traditional palaeopathology, also mean that retaining anonymity is sensible. The presence of heritable disease or genes for them identified in skeletal material should not lead living individuals to worry that they may also have the diseases or may carry them. Ethical and practical guidance covering these areas may also be needed for skeletal collections and could be based on existing genetic guidance (e.g. ISOGG, n.d.).

Dignity and respect

Finally, it is important to remember that all guidance points out the importance of retaining dignity and respect for the human remains under study. While there have been many arguments demonstrating that harm cannot be perpetuated against the deceased, as the subject of the harm does not exist (Scarre, 2013). This lack of a subject may also be important when considering the structural violence framework discussed in Chapter 9. Approaches to replacing this subject, for example by However, it cannot be denied that harm can be perpetuated to the living by disrespecting the supposed rights of the deceased. This is especially obvious in the case of disaster management where it has been argued that the care of the dead and the living are intimately connected (Woods, 2014), but I think this could be extended to the study of these identified collections. Respecting the dead, in this sense, is very much about respecting the living. Nevertheless, as discussed by Scarre (Scarre, 2013), the concept of respect as well as that of dignity when not clearly articulated raise further questions. Future guidelines need to clearly define these terms as well as consider how they can be used.

What is clear from many of these chapters is that a multidisciplinary approach is needed to working with these collections. The dangers of removing the individuals from their geographical, temporal, socio-economic, socio-cultural and socio-political contexts as well as remembering the frameworks for these within which we as researchers work, are clear: they limit our ability to correctly interpret the results and perpetuate biases within the collections. Again, these are not new ideas, the problems of working on post-mediaeval skeletons without a historian assisting with the documentary data was discussed more than two decades ago (Cox, 1995). I feel we also need to be working in an interdisciplinary context when creating guidelines and codes of practice with assistance from other fields including bioethicists, philosophers and, including anatomists in those countries where anthropology, archaeology and anatomy are disparate fields.

The aim of this introduction was to provide a background to the broad themes running through this book as well as raise questions in the minds of the reader. We are priviledged to work with these remains and we must always remember that some people disagree with their study and some vehemently wish to remain buried forever once they are dead. For these reasons (amongst others), I think it is important to maintain the highest academic and ethical standards throughout our work. Hopefully, this book will also demonstrate why these identified collections are so vital to archaeology, anthropology and forensic science (as well as other related fields), but also provoke discussion surrounding best practice and ethical standards.

References

Alemán, I., Irurita, J., Valencia, A. R., Martínez, A., López-Lázaro, S., Viciano, J., and Botella, M.C. 2012. Brief communication: The Granada osteological collection of identified infants and young children. *American Journal of Physical Anthropology* 149: 606–610. DOI: 10.1002/ajpa.22165

Alves Cardoso, F., and Henderson, C. 2013. The Categorisation of Occupation in Identified Skeletal Collections: A Source of Bias? *International Journal of Osteoarchaeology* 23: 186–196. DOI: 10.1002/oa.2285

Alves Cardoso, F., Assis, S., and Henderson, C. 2016. Exploring poverty: skeletal biology and documentary evidence in 19th–20th century Portugal. *Annals of Human Biology* 43: 102–106. DOI: 10.3109/03014460.2015.1134655

Belcastro, M.G., Bonfiglioli, B., Pedrosi, M. E., Zuppello, M., Tanganelli, V., and Mariotti, V. 2017. The history and composition of the identified human skeletal collection of the Certosa Cemetery (Bologna, Italy, 19th -20th century). *International Journal of Osteoarchaeology* DOI: 10.1002/oa.2605

Belcastro, M. G., Todero, A., Fornaciari, G., and Mariotti, V. 2011. Hyperostosis frontalis interna (HFI) and castration: the case of the famous singer Farinelli (1705–1782). *Journal of Anatomy* 219: 632–637. DOI: 10.1111/J.1469-7580.2011.01413.X

BMJ. No date. Patient confidentiality | The BMJ [online] Available from: http://www.bmj.com/about-bmj/resources-authors/forms-policies-and-checklists/patient-confidentiality (Accessed 18 September 2017)

Buckley, R., Morris, M., Appleby, J., King, T., O'Sullivan, D., and Foxhall, L. 2013. 'The king in the car park': new light on the death and burial of Richard III in the Grey Friars church, Leicester, in 1485. *Antiquity* 87: 519–538. DOI: 10.1017/S0003598X00049103

Caffell, A. C., and Holst M. 2010. *Osteological analysis the church of St Michael and St Lawrence, Fewston, North Yorkshire.* York Osteoarchaeology Ltd., York.

Caffell, A. C., Roberts, C. A., Janaway, R. C, and Wilson, A. S. 2001. Pressures on osteological collections: the importance of damage limitation. In E. Williams (ed.) *Human remains conservation, retrieval and analysis: Proceedings of a conference held in Williamsburg:* 187–197. Oxford, Archaeopress.

Cardoso, F. A. 2008. *A portrait of gender in two 19th and 20th century Portuguese populations: a palaeopathological perspective.*, Unpublished PhD thesis, University of Durham.

Cardoso, H. F. V., and Gomes, J. E. A. 2009. Trends in adult stature of peoples who inhabited the modern Portuguese territory from the Mesolithic to the late 20 th century. *International Journal of Osteoarchaeology* 19: 711–725. DOI: 10.1002/oa.991

Chi Keb, J. R., Tiesler Blos, V., Albertos González, V. M., and Ortega Muñozb, A. 2017. Documentando y contextualizando la colección esquelética del Cementerio Municipal de Xoclán, Mérida, Yucatán. *Estudios de Antropología Biológica* 17: 55–68.

Collins, M. J., Nielsen–Marsh, C. M., Hiller, J., Smith, C. I., Roberts, J. P, Prigodich, R. V., Wess, T. J., Csapò, J., Millard, A. R., and Turner–Walker, G. 2002. The survival of organic matter in bone: a review. *Archaeometry* 44: 383–394. DOI: 10.1111/1475-4754.T01-1-00071

Cox, M. 1995. A Dangerous Assumption: Anyone can be a Historian! The Lessons from Christ Church Spitalfields. In A. Herring and S. R. Saunders (eds). *Grave reflections: Portraying the past through cemetery studies:* 19–30. Toronto, Canadian Scholars' Press.

Eliopoulos, C., Lagia, A., and Manolis, S. 2007. A modern, documented human skeletal collection from Greece. *HOMO - Journal of Comparative Human Biology* 58: 221–228. DOI: 10.1016/j.jchb.2006.10.003

Facchini, F., Mariotti, V., Bonfiglioli, B., and Belcastro, M. G. 2006. Les collections ostéologiques et ostéoarchéologiques du musée d'Anthropologie de l'université de Bologne (Italie). *Bulletin archéologique de Provence* Avril: 67–70.

Henderson, C. Y., Craps, D. D., Caffell, A. C., Millard, A. R., and Gowland, R. 2013. Occupational Mobility in 19th Century Rural England: The Interpretation of Entheseal Changes. *International Journal of Osteoarchaeology* 23: 197–210. DOI: 10.1002/oa.2286

Henderson, C. Y., Mariotti, V., Santos, F., Villotte, S., and Wilczak, C. A. 2017. The new Coimbra method for recording entheseal changes and the effect of age-at-death. *BMSAP* : early view. DOI: 10.1007/s13219-017-0185-x

Herring, A., and Saunders, S. R. (eds.) 1995. *Grave reflections: Portraying the past through cemetery studies.* Toronto, Canadian Scholars' Press.

IFAA. 2011. Recommendations of good practice for the donation and study of human bodies and tissues for anatomical examination. *International Federation of Associations of Anatomists.* Available from: http://www.ifaa.net/wp-content/uploads/2017/09/IFAA-guidelines-220811.pdf

ISOGG. No date. Genetic Genealogy Standards. *International Society of Genetic Genealogy* [online] Available from: http://www.thegeneticgenealogist.com/wp-content/uploads/2015/01/Genetic-Genealogy-Standards.pdf

Lopes, C., Powell, M.L., and Santos, A. L. 2010. Syphilis and cirrhosis: a lethal combination in a XIX century individual identified from the Medical Schools Collection at the University of Coimbra (Portugal). *Memórias do Instituto Oswaldo Cruz* 105: 1050–1053. DOI: 10.1590/S0074-02762010000800016

Mariotti, V., Milella, M., Orsini, E.,Trirè, A., Ruggeri, A., Fornaciari, G., Minozzi, S., Caramella, D., Albisinni, U., Gnudi, S., Durante, S., Todero, A., Boanini, E., Rubini, K., Bigi, A., and Belcastro, M. G. 2013. Osteobiography of a 19th Century Elderly Woman with Pertrochanteric Fracture and Osteoporosis: A Multidisciplinary Approach. *Collegium Antropologicum* 37: 985–994.

Márquez-Grant, N., and Fibiger, L. (eds.) 2011. *The Routledge handbook of archaeological human remains and legislation: An international guide to laws and practice in the excavation and treatment of archaeological human remains.* London, Taylor & Francis.

Mays, S. 2016. Estimation of stature in archaeological human skeletal remains from Britain. *American Journal of Physical Anthropology 161: 646 - 655.* DOI: 10.1002/ajpa.23068

Molleson, T., and Cox, M. 1993. The Spitalfields Project: Vol. 2, the Anthropology, the Middling Sort. *CBA Research Report* 86. London, Council for British Archaeology.

Muller, J. L., Pearlstein, K.E., and de la Cova, C. 2017. Dissection and Documented Skeletal Collections: Embodiments of Legalized Inequality. In K. C. Nystrom (ed.) *The Bioarchaeology of Dissection and Autopsy in the United States*: 185–201. New York, Springer International Publishing. DOI: 10.1007/978-3-319-26836-1_9

Rissech, C., and Steadman, D. W. 2011. The demographic, socio-economic and temporal contextualisation of the Universitat Autònoma de Barcelona collection of identified human skeletons (UAB collection). *International Journal of Osteoarchaeology* 21: 313–322. DOI: 10.1002/oa.1145

Ross, A. H., and Manneschi, M. J. 2011. New identification criteria for the Chilean population: Estimation of sex and stature. *Forensic Science International* 204: 206.e1-206.e3. DOI: 10.1016/j.forsciint.2010.07.028

Salceda, S. A., Desántolo, B., Mancuso, R. G., Plischuk, M., and Inda, A. M. 2012. The 'Prof. Dr. Rómulo Lambre' Collection: An Argentinian sample of modern skeletons. *HOMO - Journal of Comparative Human Biology* 63: 275–281. DOI: 10.1016/j.jchb.2012.04.002

Santos, A. L., and Roberts, C. A. 2006. Anatomy of a serial killer: Differential diagnosis of tuberculosis based on rib lesions of adult individuals from the Coimbra identified skeletal collection, Portugal. *American Journal of Physical Anthropology* 130: 38–49. DOI: 10.1002/ajpa.20160

Scarre, G. 2013. 'Sapient Trouble-Tombs'?: Archaeologists' Moral Obligations to the Dead. In L. Nilsson Stutz and S. Tarlow (eds). *The Oxford Handbook of the Archaeology of Death and Burial:* 665-676. Oxford, Oxford University Press; DOI: 10.1093/oxfordhb/9780199569069.013.0037

Scheuer, J. L., and Bowman, J. E. 1995. Correlation of documentary and skeletal evidence in the St. Bride's crypt population. *Grave Reflections Portraying the Past through Cemetery Studies:* 49–70 Toronto, Canadian Scholar's Press.

Scheuer, J. L., and Bowman, J. E. 1994. The health of the novelist and printer Samuel Richardson (1689-1761): a correlation of documentary and skeletal evidence. *Journal of the Royal Society of Medicine* 87: 352–355.

Scorrano, G., Mazzuca, C., Valentini, F., Scano, G., Buccolieri, A., Giancane, G., Manno, D., Valli, L., Mallegni, F., and Serra, A. 2017. The tale of Henry VII: a multidisciplinary approach to determining the post-mortem practice. *Archaeological and Anthropological Sciences* 9: 1215–1222. DOI: 10.1007/s12520-016-0321-4

Ubelaker, D. H. 2014. Osteology Reference Collections. In C. Smith. *Encyclopedia of Global Archaeology:* 5632–5641. New York, Springer International Publishing. DOI: 10.1007/978-1-4419-0465-2_159

Walker, P. L. 1995. Problems of Preservation and Sexism in Sexing: Some Lessons from Historical Collections for Palaeodemographers. In S. Herring and S. R. Saunders (eds). *Grave Reflections Portraying the Past through Cemetery Studies:* 31–48. Toronto, Canadian Scholar's Press.

Woods, S. 2014. Death duty – caring for the dead in the context of disaster. *New Genetics and Society* 33: 333–347. DOI: 10.1080/14636778.2014.944260

Chapter 2

Archaeological human skeletal collections: their significance and value as an ongoing contribution to research

Jelena Bekvalac[1] and Dr Rebecca Redfern[1]

[1] Centre for Human Bioarchaeology, Museum of London

Introduction

This chapter highlights the unique large scale collections of archaeologically derived human skeletal remains curated at the Museum of London and the significant role they have played in contributing to bioarchaeology and the study of the past. The broad range of studies based upon the remains of these people who lived and died in London over the course of over 2,000 years has been widely beneficial to research and aided investigations greatly which with recent advances in science have shifted what was once considered improbable to being attainable. With the establishment of the Centre for Human Bioarchaeology (CHB) it has been able to implement curatorial standards for the long term care of the skeletal remains to support research access, contribute to discussions about the ethical issues associated with the retention of human skeletal remains and the formulation of museum human remains policy and ethical standards. The temporal range of the collections for one city through time from Prehistory to the 19th century provide an outstanding scope of research and the sustainability of such an extensive curated collection is paramount for enabling continued research of the skeletal remains allowing for new and innovative research to continue to thrive. The skeletal remains represent people from all spheres of society and highlight the diverse and cosmopolitan nature of London over time.

A small number of the curated skeletal collections are cremated remains from the Iron Age and Roman periods, but the majority are inhumations and it is amongst these individuals that a proportion from the post medieval have associated biographical information with them. The processes of time and social status has for most of the curated individuals meant that biographical details are not present for all of the collection but those without such biographical information are treated with the same care and respect, providing equally important information about the people and the past. All of the individuals within the skeletal collections would have had names but unfortunately we do not have such information for them all. The identified skeletal remains offer additional advantages for research, notably forensic studies and the opportunity for accessing documentary sources. They are a valuable component of the skeletal collections offering personal insights to these people from a wide variety of documents and accounts creating an even more detailed biography of their individual life. As with all research there are challenges and biases to consider, with human remains there are ethical issues to consider, as well as the inherent biases coming from an archaeological context and the potential limitations of the data. Overall, however, the skeletal collections provide a rich and rewarding source of information which continues to reveal a fascinating insight to the past directly through the people.

Archaeological Human Skeletal Remains Collections, Museum of London

Museums and institutions with holdings of human remains have been acquired in a variety of ways over many years, ranging from selective prospecting from different parts of the world to those that have been part of rescue archaeology. The human skeletal collections at the Museum of London (MoL) are all derived from urban rescue interventions resulting from the constant development in London and have for many years been retained as part of the archaeological archive. The current curated collection at the MoL of archaeologically derived human skeletal remains is c.20, 000 individuals (Figure 1), making it one of the largest and most significant world class collections of stratified human skeletal remains. With new development sites continuing to be excavated and the retention of more skeletal remains this number will continue to rise in the future, which in turn will also see the number of identified individuals increase.

FIGURE 1. MUSEUM OF LONDON, ROTUNDA STORE FOR THE ARCHAEOLOGICAL SKELETAL COLLECTIONS.

The Museum of London, with its skeletal collections is only one of a number of museums and universities in the UK to retain human skeletal remains that are actively sought for research. The British Museum and Natural History Museum have a diverse mix in their human remains collections, having a combination of archaeological remains from UK sites as well as individuals less than 100 years old and therefore requiring a Human Tissue Licence. The archaeological collections of human remains (skeletal and mummified) held by them collectively provide a wealth of research opportunities for comparative studies. Both hold the remains of indigenous people and The Natural History Museum has consequently worked through a number of cases with the very complex process of repatriation. Historic England as well as providing guidance for dealing with archaeological human remains also aids in supporting research and access to two uniquely important collections, the skeletal remains from the deserted medieval village of Wharram Percy (Mays, 2007) that continues to be an important focus for a range of research and the innovative curation of the large stratified non-metropolitan skeletal collection in the church of St Peter's, Barton on Humber (Waldron, 2007; Rodwell, 2011). The Biological Anthropology Research Centre (BARC), at the University of Bradford curates a number of very important and unique collections, including the victims of the Battle of Towton (Fiorato et al, 2007) and individuals excavated from a Medieval Leper Hospital in Chichester (Magilton et al, 2008). The University of Durham and Southampton also have a number of curated collections which aid in teaching and are available for research projects. Southampton University recently publishing new research on a 1,500-year-old skeleton curated from an excavation in Great Chesterford in Essex found in the 1950s, with evidence for possibly one of the earliest cases of leprosy in the UK (Inskip, 2015). Such findings many years later are an important indicator for the need to retain and curate skeletal collections to enable access for continued analysis.

With the rapidity of growth and development of urban centres there have always been challenges faced in attempting to preserve the heritage of our past. Within the latter half of the 20th century pressures increased enormously attempting to record and preserve the intrinsic remains of the past be it structural, artefacts or human remains. The disturbance of human remains in the UK whether recent or from the past requires permission for them to be disturbed and removed, which can come under the Church of England jurisdiction or secular legislation and the application of the Burial Act 1857, requiring a burial licence (Elders, 2015). The skeletal remains curated by the MoL all have a burial licence, issued from 2006 by the Ministry of Justice and prior to that the Home Office, enabling them to be retained for ongoing research. Over the course of recent years, the Ministry of Justice worked with archaeological contractors and advisory panels such as the Advisory Panel on the Archaeology of Burials in England (APABE) in the requirements and formulation for the burial licence form and issuing a licence for archaeological contractors.

The time period from which human remains may be discovered will vary from site to site, ranging from thousands, to hundreds of years and in some cases more recent times in living memory of relatives. The Human Tissue Authority (HTA) is a regulatory body for human bodies, tissues and organs with the Human Tissue Act 2004 repealing and replacing previous laws underpinned by the main principle of consent. Collections where there are soft tissues present or the remains are less than 100 years old require a licence to be issued by the HTA (Human Tissue Act, 2004). The Museum of London within its archaeological collections has only skeletal remains and all are over 100 years old. The museum abides by the Human Tissue Act 2004, which stipulates that the remains of individuals less than 100 years old may not to be retained without a licence. The foul conditions associated with 19th century burial grounds, due to high levels of overcrowding, led to parliament passing a series of Burial Acts in the 1850's leading to the closure of numerous London burial grounds and crypts. For the most part, new commercial developments within the City of London, although having the potential of disturbing such areas, would be responsible for the majority of human remains findings, more than 100 years old.

The archaeological human skeletal collections have become a unique research archive in learning about the people as individuals and multi-faceted aspects of the past directly through them. The impact of them as a source of understanding the past is enhanced further due to their stratified nature providing a temporal span of 2,000 years for one city coupled with the associated archaeological contextual record. This provides details of location, placement and treatment of the body, grave type, artefacts, relation with other burials and subsequent disturbances. The many transitions of London's history are expressed through the tangible link these people have to the times in which they lived, worked and died, making them a fascinating focus for research. The human skeletal collections at the museum have long been the first port of call for carrying out research with that trend increasing over the years as the fields of osteology and bioarchaeology have developed (Roberts, 2006) and with the ongoing retention of varied and large collections. Pivotal to making the MoL collections such a landmark in the field of research was the establishment of the Centre for Human Bioarchaeology at the Museum of London, as a part of the London Archaeological Archive Research Centre (LAARC).

Centre for Human Bioarchaeology

Prior to the existence of the Centre for Human Bioarchaeology (CHB) at the Museum of London the ability to support access for research to the skeletal remains was limited. The advent of the CHB enabled staff to manage and support access to the collections and ultimately freely share the osteological data on line from 2007, which had a marked effect on the focus and output of osteological research. The CHB effectively increased the awareness of the skeletal collections making the collections and data more readily accessible at a national and international level, exponentially increasing their significance and importance for comparative research.

The CHB was established with funding from the Wellcome Trust in 2003 with a team of osteologists based at the MoL analysing and recording the skeletal remains from earlier deposited collections and working in tandem with the osteologists from Museum of London Archaeology (MoLA) on the developer funded Spitalfields Market project. The Spitalfields excavations (1998 – 2001) revealed the remains of over 14,000 skeletons from the medieval monastic site of St Mary Spital and with the retention of 10,500 skeletons is the largest single stratified human skeletal collection (Connell et al. 2012; Thomas, 2004). The two teams worked over a period of three years to enter the osteological data in to a bespoke database (Wellcome Osteological Research Database, WORD) created by Brain Connell and Dr Peter Rauxloh. The Oracle platform relational database enabled rapid recording of the skeletal remains (Connell, 2003 and Powers, 2012) and followed standard osteological methods of recording (Brickley and McKinley 2004; Buikstra and Ubelaker, 1994). With the large scale of the skeletal collections and specifically the extensive number retained from St Mary Spital it was necessary to have a means of capturing the data effectively and efficiently, whilst enabling the data to be queried and provide a usable platform to add further collections in conjunction with the long term sustainable access to the data.

The benefits of the database are numerous, allowing digital access to osteological data in a way not previously possible, linking with the archaeological context and the ability for interrogation of large datasets. As a research mechanism the database is a formidable tool able to carry out complex data queries that are intrinsically linked with the archaeological data fields to produce searchable datasets in an Excel sheet format. The database is a dynamic research engine which is continually being added to with new skeletal collections recorded as well as updating existing records to show other applications that may have been undertaken on the skeletal remains, such as destructive sampling and radiography. The ability to run queries on the data produces large quantitative datasets for statistical analysis and enables the opportunity to address small and large scale research questions with relative ease. It has proved to be invaluable for research and as a curatorial tool for aiding in the long term conservation of the skeletal collections.

The archaeological context for research is imperative, as too is the consideration of the context for the human skeletal remains retained and curated being the remains of once living people. As such they have a singular sensitivity to them and the working practices around them as a means for research and display need to be accountable and robust. When the CHB took up the mantle of the curation of the archaeologically derived human skeletal remains in 2003, it was closely involved in the development of frameworks and guidelines surrounding the complex and diverse issues of ethics and standards. The ethical treatment of human remains became a topic of much discussion and debate, particularly in light of the issues surrounding repatriation cases. Hedley Swain, the former head of the archaeological collections at the Museum of London, was a key figure in the discussions as to how human remains in museums were to be managed and was the Chair of the Department for Culture, Media and Sport (DCMS) Working Group which led to the formulation and publication of the DCMS *Guidance for the Care of Human Remains in Museums* (2005).

Leading on from this the Museum of London was the first museum in the British Isles to produce a Human Remains Policy (Museum of London, 2006) revised in 2011 (Museum of London, 2011) outlining the museum's approach to the care, handling, access, research, display and ethics of retaining human skeletal remains.

'There is an ongoing debate as to the ethics of excavating, holding and displaying human skeletons by museums. This is a complex and multi-layered debate, influenced by concerns of Indigenous peoples in other countries; the multi-cultural nature of modern society; as well as modern religious and humanist philosophies, medical ethics and museological concerns.' (Museum of London Human Remains Policy, 2011 page 3)

In 2004 the Museum of London hosted an international conference at the Museum of London, Docklands and produced a published volume which focused on museum practice with particular attention to the collection of human remains, and their display where 'the purpose was to draw out different perspectives that influence both the policy framework and museum practices' (Lohman and Goodnow, 2006: 15). Curating such a large collection of skeletal material, MoL was well positioned to host such a conference. It had previously held gallery displays of human remains, culminating in the production of the London Bodies exhibition in 1998, and yet it remained a neutral host for the more controversial repatriation cases. The discussions were lively and at times contentious, demonstrating clearly the emotive power of human skeletal remains and the varying attitudes to how they should be dealt with when disturbed.

The CHB continues to be involved in the formulation and application of polices and practice, contributing a chapter Museum of London: An Overview of Policies and Practice to the edited volume Curating Human Remains, Caring for the Dead in the United Kingdom (Redfern and Bekvalac, 2013) and with Dr Rebecca Redfern a member of the British Association for Biological Anthropology and Osteoarchaeology (BABAO) Working Group which worked extremely hard in creating and producing for BABAO a Code of Ethics and Code of Practice for those working with archaeological human remains. The code outlines guiding principles for the ethical considerations of handling, retention and research of human remains from archaeological contexts.

Research

The primary reason for the retention and curation of the human skeletal remains from the archaeological excavations in the City of London and Greater London Area are for research and this is central to the role of the CHB. A remit of the Wellcome funded project was to make the collections accessible and to freely share information. In 2007 the CHB launched its website allowing for open access to the skeletal collections for researchers, providing information with cemetery summaries of sites recorded and the osteological data downloads for those sites to be downloaded free of charge. Such digital access to an osteological collection was unprecedented and over the ensuing years the momentum of researchers downloading the data to use remotely has grown and researchers visiting the Centre increased. In conjunction with the launch of the website, the CHB was able to actively participate and disseminate information about the database and available data for research projects through talks, study days and conferences, including those held by the American Association of Physical Anthropologists (AAPA), Paleopathology Association (PPA) and British Association for Biological Anthropology and Osteoarchaeology (BABAO), all of which added to the increasing awareness of its presence and role to assist and encourage research.

The MoL archaeological skeletal collections have become an integral component of analysis for researchers either being the sole area of their study or as part of a wider comparative study. They have proved to be so valuable because of a wide range of factors with their extensive numbers allowing for large quantitative studies with statistical significance, broad pathological profile, the different types of burials and cemeteries, identified individuals, varied geographical locations across London and a temporal range spanning over 2,000 years.

The CHB is fortunate to have been able to develop over the years relationships and connections with numerous universities and institutions in the UK and overseas, enabling a wide variety of research projects undertaken by undergraduate, Masters and PhD students, post-doctoral and collaborative research projects (internal and external) both inter- and multi-disciplinary. Supporting and managing the access to the collections with an application process for visiting researchers or using the data remotely is beneficial in enabling the CHB to chart the studies undertaken on the collections and in so doing can act as an aid in supporting other studies. For those researchers only remotely accessing and using the digital data downloads, a 'User Data Form' is completed so that the Centre knows who is carrying out the study, what it entails and at which institution they are based. With the data downloads and visits to the Centre it allows for the CHB to see which sites are those most frequently requested enabling the CHB to identify how it can best support the research topics as well as preventing the over handling of the skeletal remains and repetition of studies. Additionally, for the skeletal remains directly accessed by researchers it is a useful aid for working with the MoL conservation team for implementations in the long term curation of the skeletal remains.

Year on year since the launch of the website in 2007 the number of researchers using the collections has increased. This was explicitly shown when as part of an Institute of Field Archaeologists (IFA) Seminar in 2013 held at the LAARC in Mortimer Wheeler House, Hackney to review research and access to archaeological collections, the CHB was invited to contribute. A snapshot for the period from November 2012 to November 2013, in relation to users of the CHB website and its pages, showed with figures provided by the information technology team the high number of visits and usage of the website

- Visits: 14,431
- Visitors: 12,124
- Page views: 62,707

For that same period of time there were 60 researchers that visited the CHB to collect data directly from the skeletal collections and there were 31 submissions of the 'User Data Form'. The number of researchers using and accessing the skeletal collections has consistently remained with the same high numbers.

The application process also enables the CHB to have a clear indication of the proposed study by the researcher, whereby it can advise, when needed, with the focus of the study and aid in the selection of the sites and individuals. This also aids in implementing a long term research strategy for proposed projects to assist in addressing questions in areas where there are gaps in the information, which is particularly pertinent in combination with the increasing developments in destructive applications for them to support a well-defined research strategy. The CHB formulated a Destructive Sampling Policy in response to the increasing requests for research projects with destructive tests at the core of the study. Such studies have the potential for answering questions not possible with macroscopic analysis alone and enhancing the osteological narrative. However, the consideration of the integrity of the skeletal remains is paramount for the sustained curation of the skeletal remains, making it imperative to have a robust policy for the sampling application process with a collective decision procedure which at the MoL is overseen by the Collections Committee.

The Destructive Sampling Policy sits in association with the earlier construct of an Archaeology Research Framework (Nixon et al 2002), which at that time set about producing a framework for London archaeology with three aims; 'realising the potential of the London Archaeological Archive, managing the archaeological resource more effectively and facilitating better focused research' which alongside had a series of research priorities that included people and society. As a consequence of the creation of the Oracle platform database (WORD), implementing the analysis and recording of the human skeletal remains with the subsequent access for research to the skeletal remains, the CHB has been able to achieve the three principles of the framework and address answering the questions about the people and society demonstrated with the wealth of research output. It is hoped that the framework will be updated in the future in light of the advances since its formulation and new scientific developments for application with archaeological collections.

Research Output

The dissemination of information from analysis of the skeletal collections is vital to maintaining the awareness of the collections, highlighting their significance and validity and sharing with other researchers the scope of the available data. All such outputs add enormously to the knowledge of the collections and are invaluable for the continued contribution of information to the museum and the curators of the skeletal collections. The CHB website, conference proceedings, student theses, journal and book publications have produced a wealth of invaluable and ground breaking research results from analyses incorporating the MoL collections. To list them all would not be feasible but the breadth and scope of studies has been all encompassing ranging from studies learning about particular diseases, diet and the environment (Spencer, 2008; Bernofsky, 2009), the effects and implications of metabolic disorders (Ives, 2007; Brickley and Ives, 2008; Ives *et al.* 2016), growth and development (Watts, 2013), and anatomical morphology (Plomp, 2015). Studies have focused on specific time periods, including explorations of the lives of those living in the Roman world with investigations into puberty (Arthur et al, 2016), mobility and diet, (Redfern 2016) and the impact on the health of individuals living in London during the Roman era and identifying migrants coming to London (Shaw et al, 2016). The tumultuous times of the medieval period have equally revealed how these events are reflected in the markers of health faced in often hazardous environments (Antoine et al, 2005, Antoine and Hillson, 2005; Yaussy, 2016), and the implications of this seen in oral health (De Witte and Bekvalac, 2010;) and bony lesions as markers of frailty (De Witte and Bekvalac, 2011), trauma in children (Verlinden, 2015) and dental health (Walter et al, 2016). Insights to health variance of those in different monastic orders (Redfern and Bekvalac, 2014) and mortality comparison for those in monastic enclaves compared to the lay population (De Witte et al, 2013). The perils of the post medieval period and effects on child health from the dentition (Hassett, 2011a, 2011b), diet and dental health (Mant and Roberts, 2015) the effects of body size to mortality (Hughes-Morey, 2012) and the stark contrast of social strata in the transition to urbanisation with markedly detrimental effects to health and mortality (De Witte et al, 2016), and the impacts to childhood growth and health in post medieval England (Newman 2015; Newman and Gowland, 2016). Developments in the sampling of teeth for stable isotope analysis have proved critical to more accurately charting an individual's life, looking at signatures of diet, mobility, migrants and famine (Kendall, 2013; Beaumont et al, 2013a & b) and the fundamental progress in ancient DNA analysis for identifying the actual pathogens of diseases (Schuenemann et al, 2013).

Two recent doctoral studies accessing the skeletal collections utilised an innovative incorporation of contemporary documentary sources for compelling comparisons with the skeletal record. The Medieval Court roles from London were scrutinised for comparison with incidences and type of trauma in the skeletal record from the extensive medieval collections by Kathryn Krakowka, revealed a beguiling and previously unexplored avenue for a better perception of life in London at that time (Krakowka, 2015). The documentary sources for the post medieval period become more abundant and are a very rich source for comparison with 18th and 19th century skeletal remains curated from cemeteries and crypts. The post medieval collections at the CHB cover a broad spectrum of locations and types of burials reflecting London at a time of urban change on an unprecedented scale and in many instances have individuals with associated biographical details. The archives of hospital admittance records have proved to be an interesting source of information and Madeline Mant was able to access a variety of hospital records for the Royal London and St Bartholomews hospital for comparison with individuals from the post medieval collections. The ensuing research and results produced a fascinating insight for the time and people treated, and highlighting to some of the problems faced with using such historical datasets (Mant, 2016).

The scale of the collections allows for support of different projects, from small focused studies to large scale population projects addressing bigger questions about health, mobility, diet, growth and development. Two such large scale research projects, in which the inclusion of the collections was a significant component, included the University of Durham Natural Environment Research Council (NERC) funded Global tuberculosis (TB) Health project, an ancient DNA study of tuberculosis from prehistory to the post-Medieval period in Britain and parts of Europe and the University of Reading Leverhulme funded project Adolescence, Migration and Health in Medieval England: the osteological evidence. The MoL collections were able to contribute to the tuberculosis study with samples to aid in the examinations of the evolution and development of a disease that has had an enormous impact on the lives of humans (Müller et al, 2014). The adolescence project was able to investigate this area in a way that had not previously been possible, producing   interesting and significant results leading to a methodological approach to aid in identifying pubertal growth (Shapland, 2013; Lewis et al, 2016a, 2016b).

Publications by the Museum of London Archaeology (MoLA) have for many years produced study series and monographs on the London excavations providing an essential source of reference material with osteological and archaeological data from the many cemetery excavations. These range from the Roman period with the Eastern cemetery excavations (Barber and Bowsher, 2000), the Royal Mint excavations revealing the Black Death Catastrophe cemetery at East Smithfield (Grainger et al, 2008) and St Mary Graces Abbey (Grainger and Phillpotts, 2011), and the influential bioarchaeological work on the medieval St Mary Spital excavations (Connell et al, 2012). The often dire conditions experienced by those living and dying in the poorest parts of London (Brickley and Miles, 1999) and with more recent developments revealing a high number of post medieval burial grounds, adding to this vision of a population under stress at a time of great change drawing upon the bioarchaeological construct to produce engrossing historical narratives (Miles et al, 2008a; Fowler and Powers, 2012; Henderson et al, 2013).

All research undertaken on the skeletal collections (published and unpublished) is valuable, each time adding another layer of information and knowledge about them. As part of the archive at the CHB it holds copies of student research on the collections and asks, as part of the access, for this to be deposited by the student. The grey literature of unpublished skeletal reports is large and a valuable asset that is not always readily accessible to other researchers and is an area that needs to be addressed. BABAO has been undertaking work towards a means of opening up access to the grey literature and realising its potential.

Destructive Sampling

The developments in chemical analysis and destructive sampling have moved at a considerable pace since the opening of the CHB and as such had a profound effect on the potential of what can be revealed from the archaeological skeletal remains. The fields of ancient DNA and stable isotope analysis have been able to provide mechanisms for answering questions that in the past could only be imagined. One such ground breaking piece of research from the MoL collections was that of the findings from the research for a PhD on the East Smithfield Black Death Catastrophe Cemetery. This was a collaborative project with McMasters University, Canada, Dr Sharon DeWitte and the CHB. A number of individuals were sampled from the cemetery by Dr Sharon De Witte and Kirsten Bos for her PhD with the objective of being able to identify the causative agent of the Black Death, a longstanding issue of contention (Figure 2). There had been much debate about the actual cause of the disease which killed such vast numbers of people, having a devastating impact on the UK and the world as whole with social and political implications that were felt many years after the disease had taken such

FIGURE 2. DR SHARON DE WITTE AND KIRSTEN BOS (AT TIME PHD STUDENT) TAKING SAMPLES FOR ANCIENT DNA ANALYSIS OF INDIVIDUALS FROM EAST SMITHFIELD BLACK DEATH CEMETERY.

a horrible toll on the population in 1348 -1350. The resultant data that was achieved from the samples was ground breaking and led to the confirmation that the causative agent of the Black Death was *Yersinia pestis* and from the integral work by Bos and McMasters it was possible for the first time to reconstruct an ancient pathogen genome (Schuenemann et al, 2011; Bos et al, 2011). The ramifications of such findings are far reaching and will continue as the developments for ancient DNA move on further and the investigations of our ancestors are increasingly possible to look at from a genetic perspective. The 2015 Written in Bone exhibition at the MoL was able to showcase some of the fascinating research led by Dr Rebecca Redfern utilising the new aDNA research to reveal hidden insights to the lives of the first Londoners (Redfern et al 2017). Providing genetic details about them with ancestral origin, hair and eye colour, unique chromosomal information and markers of pathogens not manifest in observable bony lesions observable.

Biographical collections

Research on the MoL skeletal collections have also contributed to studies in the fields of art, medicine (Naveed et al, 2012) and forensics. Forensic researchers can use the named individuals in the collection to test theories against biographical information provided by the coffin plates associated with the skeletal remains. Knowing the sex and age of the individuals is a key element for forensic research projects and provides a stand point from which methods can be tested and developed with a high level of confidence.

These can prove invaluable for being able to refine and improve methods and test new methods. Skeletal collections where there are identified remains predominantly come from excavations of sites with remains from the post medieval periods and more often associated with individuals of middle to high social status. The choice and method of burial or interment in a crypt of these individuals most often in triple shell lead coffins with lead coffin plates (inner and depositum) increases the likelihood of the biographical information recorded to survive.

The post medieval collections at the CHB have a large sample of known named individuals from those revealed from excavations at Chelsea Old Church (25), (Cowie et al, 2008) St Benet Sherehog (5) (Miles et al, 2008a), Bow Baptist burial ground (45), (Henderson et al, 2013) and the recently deposited archive from excavations of a playground at St Johns School, Bethnal Green (360) by AOC Archaeology with a publication of the site pending. With recent excavations of post medieval sites and future large scale engineering projects such as Crossrail and HS2 the number of identified individuals will be greatly increased.

The analysis of the skeletal remains whether having biographical detials or not are all recorded in the same way into the digital database but to indicate that there are additional details the database has a tick box for highlighting that biographical details are known. The name of the person if known is noted in the comments box with the birth and death dates given with the age at death. This information is also all included in the downloadable data from the Centre for Human Bioarchaeology, Museum of London website for the respective sites which have identified individuals. The identified individuals all have context numbers assigned to them in the same way as the other individuals excavated from the sites and are able to be searched for in the database utilising the sitecode and context number. This enables studies that may wish to include them to carry out blind tests as the data can be filtered accordingly so that the biographical details are not immediately visible. The skeletal remains of the known named individuals are curated and stored following the same guidelines as with all of the archaeologically derived skeletal remains retained in the Museum of London. The one variation being with the identified individuals from the excavations of the crypt in St Bride's being retained within the church. They are an invaluable reference collection on which a large range of studies have been carried out for the testing and formulation of applications and methods. The additional information of knowing the age and sex for the individuals provides the confidence needed for comparative analysis when testing applications and methods. However, there should also be a degree of caution as there can in some instances sometimes be the potential for there to be an incorrect association between the skeletal remains and coffin plate.

The CHB also assists with the curation of the 227 known named individuals at St Bride's Church, Fleet Street (Scheuer, 1998) which are contemporary with the known named individuals from Christ Church Spitalfields curated at the Natural History Museum, London (Molleson et al, 1993, Reeve and Adams, 1993) which together are uniquely important collections for comparative research. Improving the means for sex and age estimation is important in osteoarchaeology for demographic accuracy but an imperative in forensics with the legal implications for proof of identity. The skeletal collections with biographical details for individuals are a vital component to being able to successfully approach the advancement of forensic techniques and methods. The 18th and 19th century skeletal individuals from the crypt at St Bride's church, revealed during archaeological excavations of the church in the 1950's by Professor Grimes from the aftermath of WWII bombing, have been central to a number of earlier forensic and osteological studies (Steel, 1960; Berry, 1976).

Since 2008 the CHB has assisted the church with the curation and research access to the named individuals, who continue to aid research with methods employed for sex estimation (Gapert et al, 2009a; Gapert et al, 2009b) and the recent recording of them for aiding in the refining of the transitional aging technique, which has proved to have a greater degree of accuracy for aging adult individuals in archaeological collections which predominantly have no biographical information. Collaboration with the Cranfield Forensic Institute begun in 2012 used 3D scanning equipment in the crypt taking scans of the pelves and mandible of individuals for the Virtual Skeletal Analysis (ViSA) project for the development of methods to aid in age estimation. In 2016 digital orthopantogram radiographs were taken of selected females and males using dental equipment at the Royal London Hospital (Figure 3). A small number were scanned using a cone computerised tomography (CT) scanner with the resultant data for a master's thesis to be analysed as an aid in establishing a means of dental identification for forensic cases.

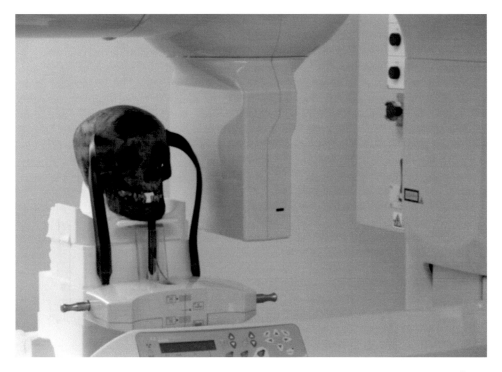

FIGURE 3. ROYAL LONDON HOSPITAL, ORTHOPANTOGRAM OF FEMALE INDIVIDUAL FROM ST BRIDE'S.

The opportunity for being involved in such a wide and all-embracing array of fields for research enhances the knowledge of the collections, disseminated in many formats to engage academics and the public. The identified collections play a major part in such studies in a twofold way being able to be incorporated in regard to being part of the skeletal analysis but also with the added benefit for different strands of research that need the biographical data as an integral feature of the study. At the same time actively showing that the collections are consistently used, curating them is necessary so that new research can be applied and demonstrate that they continue to be of relevance for broadening the scope and horizons of research. All institutions face challenges for supporting the long term sustainability of its holdings and as such it is imperative that collections continue to be utilised and accessed, with the resultant research open to the public domain. Fortunately the archaeological skeletal collections curated by the CHB are able to show that they are an active and dynamic source of research with measurable results from public engagements and the output from researchers using the collections either solely or as part of their research projects.

Outreach, Teaching and Exhibitions

Sharing information with the public audience is another important facet of the CHB to promote the museum in achieving its strategic aims and showcase the research on the collections. As the CHB has established itself within the museum it has become more actively involved in public facing outreach events and exhibitions. Being able to directly participate in public events is a principle means of engaging with a public audience and a means of being able to engage them face to face to provide them with a much better awareness of how and why it is that the museum has such an extensive collection of human skeletal remains. Human remains have an emotive association for people and these events are a means of sharing information about how the skeletal remains are retrieved within an archaeological process in response to the consequence of the building developments in the City of London and Greater London Area. A discussion is possible for there to be a better understanding that disturbing the remains of these people is not at the instigation of the osteologists but the result of the constantly changing landscape of London, and that with that disturbance of these people they offer a unique opportunity for research to learn about them, the past and the present.

Engaging with the public in this way allows for conservations about the ethics and issues around curating human skeletal remains and for their voices and opinions to be heard. More often than not it is the first time that the visitors have come in to contact with human remains and seen a human skeleton. It offers a unique experience for them and approach of learning about the past from the actual people who once lived in London. The circumstance in which to be able to answer their questions about the skeletal remains is invaluable, to discuss what research has taken place and is on-going allows the CHB to highlight the value of the research work and allay any potential concerns they may have. The outreach events also enable a younger audience to be able to participate and the inquisitiveness of the young is refreshing and can be challenging as they always want to know more. Having the opportunity to interact with the younger audience at museum open days, Festival of Archaeology, After School Club and Young

Osteology Group enables them to learn about the value of the archaeological collections which is important as they are hypothetically our researchers of the future.

The CHB is very fortunate to have a large and varied teaching collection consisting of disarticulated skeletal elements (adult and non-adult) retrieved from earlier excavations that have excellent examples of pathologies and trauma, in addition to mixed anatomical elements and crania and pelves for the use of estimating sex and age. The dedicated teaching collection is an invaluable means of actually being able to show the effects of pathology on the bones and teeth. It was instrumental in being able to launch the adult evening teaching course that was first instigated by Lynne Cowal, a Wellcome Team osteologist, which continued with the Bare Bones Course that ran until 2014 when the museum ceased to have an in house adult learning coordinator. The adult teaching courses were always popular and had an assorted mix of participants with different backgrounds and interests in the past. Student study days can also be focused around particular topics as requested by the university or school to help with providing a greater insight to anatomy and pathology

Exhibitions are a wonderful way for reaching a diverse audience and to convey from research on the collections an enthralling and informative narrative directly from the people. The first large scale exhibition at the museum which had human skeletal remains at its core was the very well received 'London Bodies: the Changing Shape of Londoners from Prehistoric Times to the Present Day' in 1998, which brought together skeletal individuals from different time periods and using methods of analysis at that time including facial reconstruction to highlight was possible to learn from these people living in the past (Werner, 1998). Twenty years later in 2008 Skeletons: London's Buried Bones was opened at the Wellcome Collection, made up of 26 skeletons selected by the team at the CHB from the curated collection. Those selected came from cemeteries across London from the Roman, Medieval and Post Medieval periods to show what lies beneath an ever changing city and how they expressed through their skeletal remains the impacts to their health and well-being when alive. Within the selected individuals for display in the exhibition there were three individuals from Chelsea Old Church where they were known named individuals. A search was made to see if any relatives carrying out research on them and in the area but there were none at that time. The decision to include these particular individuals was that they had such interesting associated narratives about them from being able to go to other documentary sources. There were a number of discussions as to whether those with known named should be included in the exhibition but the decision was finally made to include them for demonstrating how more can sometimes be known about the individuals and that can further enrich the narrative. Based at the Museum of London in 2012 Doctors, Dissection and Resurrection Men was an exhibition formulated around the remarkable findings from the excavations with the developments at the Royal London Hospital and provide a snapshot of a critical point in medical history, when the cemetery revealed covered the period from 1825 – 1841, with the critical Anatomy Act 1832 passed during its usage (Fowler and Powers, 2012). A regional tour based on the original Skeletons exhibition will over a period of two years (2016 – 2018) be based in three different locations collaborating with regional museums and having skeletal remains form the environs of those areas for a comparison to the archaeological remains from London.

Conclusion

The identified skeletal collections are an important group within the curated collections and are often an integral inclusion in studies and research because of having the associated biographical information. Primarily it is the middle to high status individuals which make up the identified collections from the 18th and 19th century relating to the preservation and legibility of the coffin plate, which does then have a bias for such associated information to relate predominantly to the higher status groups within the collections. A question of whether they should be treated in a different way to the other skeletal remains is asked but in terms of how they are recorded and cared for there is not a differentiated treatment of them based upon having such biographical detail. The biographical detials are a valuable source and make the individuals in such a collection as it were 'special' but more so in terms of how the biographical data can be an advantageous aid in with respect to particular research projects.

 An ever increasing interest with on line research in to family histories and genealogical studies has led to a number of enquiries in relation to the named individuals curated at the CHB. With the data available online for sites with identified individuals those carrying out family research have more often found that potentially one of their relatives from the past is no longer in the place they were buried but retained within a museum. For those who have contacted the CHB they have always been fascinated by the information which is able to be imparted about their distant relative. Such enquiries are always dealt with respect and mindfulness that the skeletal remains have an added poignancy to the person enquiring in having a familial link. They could if they wished to request with the correct and necessary paperwork to remove the individual form the archive for them to be reburied, for those who have contacted MoL this has not so far been a requested made from a family member. The majority of enquiries have come from relatives with family buried at Chelsea Old Church and St Bride's and all of those who have contacted the CHB have been extremely generous in the family information that they have then be willing to share to enable an even better insight to the person's life and answer question .

The nature of archaeologically derived skeletal collections is inherently different from those that have been collected with a specific purpose and focus such as medical pathological collections. With all collections there are limitations and biases and archaeological human skeletal collections are not exempt the large size of the MoL collections does enable the chance to militate against some of the problems faced with archaeologically derived material but there will always be imperfections and biases in the samples. The human skeletal remains from excavations are beset with the problems of bias and consideration of this must be taken in to account when carrying out research with them. The identified individuals within the archaeologically derived collections also have similar issues of bias but these can often be based more upon other factors such as the degree of soft tissue preservation in regard to the manner in which they had been buried or interred. Primarily, the commercially driven excavations from which the majority of archaeological skeletal collections are derived have limitations with time and money being funded by the developer. The area that is to be developed will not be entirely excavated and so what is in effect already a sample of a cemetery or burial ground is sub-sampled further for the lifting, retrieval and analysis of the skeletal remains. It is important to be aware of sampling and to consider such sampling strategies for the potential implications this will have for the data and subsequent analytical results with research of a collection. There will be limitations and constraints on what may be feasible and possible which have to be taken in to account when accessing and utilising the archaeological collections and those with biographical information.

The output of research from archaeological human remains is vast and all-encompassing with the MoL collections having a significant input and impact to the resultant data. The large scale collections from London excavations are a remarkable resource for research, but being readily available on a digital basis has had a positive and negative effect in making them more accessible for a broader audience which can then have the effect of skewing results creating a bias. This effect with the London data was highlighted in the article by Roberts and Mays, 2011, investigating the study of curated collections and future curation of skeletal collections in the UK. It is important to be aware of such bias and to encourage the access and support of other archaeological collections for research and comparative studies.

The CHB is able to demonstrate categorically the significance and value of archaeological human skeletal collections from the contribution that the MoL collections have already made to a plethora of research areas and will continue to have a valuable contribution to future research. An important aspect in the curatorial role is to demonstrate the active use of collections and emphasise how by having long term curation they are available for the application of new techniques and methods of research to answer questions about the past that resonate to the present day. The continuing investigations employing aDNA to the collections will further unlock secrets held within the bones and with the recently more accessible application of digital radiography (Figure 4) enable a means for looking inside the bones (Bekvalac, 2012 and Western and Bekvalac, 2015). The Impact of Industrialisation on London Health project, a three year funded project by City of London Archaeological Trust (CoLAT) Rosemary Green award by accessing a large number of archaeological remains, from London and non-metropolitan areas to provide evidence of health patterns have changed over time and the role that industrialisation has played in determining the factors critical to the health of London's population, past and present.

The existence of such extensive archaeological skeletal collections and digital archive at the Museum of London, are a testament to the amount that has so far been achieved in a spectrum of fields and which will be built upon as there continue to be advancements in applications for analysis. Some of the limitations and problems encountered by the authors Roberts and Cox in collating and writing the comprehensive publication Health and Disease in Britain From Prehistory to the Present Day (Roberts and Cox,2003) have been addressed and will continue to improve. In a very literal sense what has subsequently so far been achieved with improved access to archaeological remains and notably the MoL collections does in a literal sense speak volumes as to the greater cohesive research attained. The identified skeletal remains have the additional value

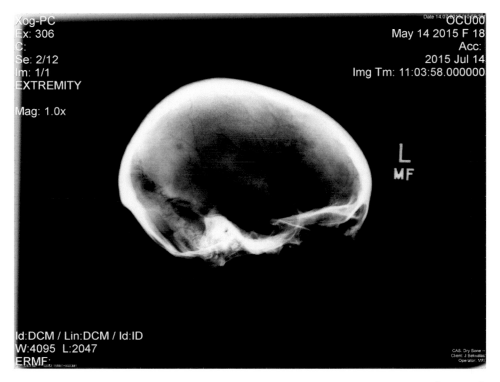

*FIGURE 4. DIGITAL RADIOGRAPHY OF THE CRANIUM OF A FEMALE FROM CHELSEA OLD CHURCH (OCU00 18)
TAKEN FOR IMPACT OF INDUSTRIALISATION ON LONDON HEALTH PROJECT*

for being known individuals and all that can be researched based on such knowledge with access to documentary sources enabling an even greater insight and knowledge of these individuals. However, irrespective if they are identified or not all of the human skeletal remains should be and are treated with care and respect. Archaeological collections of human skeletal remains and identified collections are without doubt unparalleled in the many ways in which they can inform us about the past, present and potentially future with a tangible significance for their continuing contribution to bioarchaeological research.

References

Antoine, D. M. and Hillson, S. W. 2005. Famine, Black Death and health in fourteenth century London. *Archaeology International* 8: 26–28.

Antoine, D. M., Hillson, S. W., Keene, D., Dean, M. C. and Milne, G. 2005. Using growth standards in teeth from victims of the Black Death to investigate effects of the Great Famine (1315-1317). *American Journal of Physical Anthropology Supplement* 125.

Arthur, N., Gowland, R. L. and Redfern, R. C. 2016. Coming of age in Roman Britain: osteological evidence for the age of puberty. *American Journal of Physical Anthropology* 159: 698–713.

Barber, B. and Bowsher, D. 2000. *The Eastern Cemetery of Roman London Excavations 1983–1990*. Monograph 4. London, Museum of London Archaeology.

Beaumont, J., Geber, J., Powers, N., Wilson, A., Lee-Thorp, J. and Montgomery, J. 2013a. Victims and survivors: stable isotopes used to identify migrants from the Great Irish Famine to 19th century London. *American Journal of Physical Anthropology*. 150(1): 87–98.

Beaumont, J., Gledhill, A., Lee-Thorp, J. and Montgomery, J. 2013b. Childhood diet: a closer examination of the evidence from dental tissues using stable isotope analysis of incremental human dentine. *Archaeometry* 55(2): 277–295.

Bekvalac, J. 2012. Implementation of preliminary digital radiographic examination in the confines of the crypt of St Bride's Church, Fleet Street, London. Proceedings of the Twelfth Annual Conference of the British Association for Biological Anthropology and Osteoarchaeology. *British Archaeological Reports International Series* 2380: 111–118.

Bernofsky, K. S. 2010. Respiratory Health in the Past: A Bioarchaeological Study of Chronic Maxillary Sinusitis and Rib Periostitis from the Iron Age to the Post Medieval Period in Southern England. Unpublished PhD thesis, University of Durham.

Berry, A. C. and Berry, R. J. 1976. Epigenetic variation in the human cranium. *Journal of Anatomy* 101: 361–379.

Bos, K. I., Schuenemann, V. J., Golding, B., Burbano, H. A., Waglechner, N., Coombe, B. K., McPhee, J., DeWitte, S. N., Meyer, M., Schmedes, S., Wood, J., Earn, D. J. D., Herring, D. A., Bauer, P., Poinar, H. N. and Krause, J. 2011. A draft genome of Yersinia pestis from victims of the Black Death. *Nature* 478: 506–510.

Brickley, M., Miles, A. and Stainer, H. 1999. The Cross Bones burial ground, Redcross Way, Southwark, London: archaeological excavations (1991–1998) for the London Underground Limited Jubilee Line Extension Project. Monograph 3. London, Museum of London Archaeology Service.

Brickley, M. and McKinley, J. 2004. *Guidelines to the Standards for Recording Human Remains. IFA Paper No. 7*. Reading and Southampton, Institute of Field Archaeologists and British Association for Biological Anthropology and Osteoarchaeology.

Brickley, M. and Ives, R. 2008. The bioarchaeology of metabolic bone disease. London, Academic Press.

Buikstra, J. E. and Ubelaker, D. H. (eds)1994. *Standards for Data Collection from Human Skeletal Remains. No. 44*. Arkansas, Arkansas Archaeological Survey Research Series.

Code of Ethics, 2010. British Association for Biological Anthropology and Osteoarchaeology. Available from: http://www.babao.org.uk/publications/ethics-and-standards/

Connell, B. and Rauxloh, P. 2003. *A Rapid Method for Recording Human Skeletal Data*. London, Museum of London Report.

Connell, B., Gray Jones, A., Redfern, R. and Walker, D. 2012. *A Bioarchaeological Study of Medieval Burials on the Site of St Mary Spital: Excavations at Spitalfields Market, London E1, 1991-2007*. Monograph Series 60. London, Museum of London Archaeology.

Cowie, R., Bekvalac, J. and Kausmally, T. 2008. *Late 17th to 19th Century Burial and Earlier Occupation at All Saints, Chelsea Old Church, Royal Borough of Kensington and Chelsea*. Archaeology Study Series 18. London, Museum of London Archaeology.

De Witte, S. and Bekvalac, J. 2010. Oral health and frailty in the Medieval English cemetery of St Mary Graces. *American Journal of Physical Anthropology* 142(3): 341–354.

De Witte, S. and Bekvalac, J. 2011. The association between periodontal disease and periosteal lesions in the St. Mary Graces cemetery, London, England A.D. *American Journal of Physical Anthropology* 146(4): 609–618.

De Witte, S. N., Boulware, J. C. and Redfern, R. C. 2013. Medieval monastic mortality: hazard analysis of mortality differences between monastic and nonmonastic cemeteries in England. *American Journal of Physical Anthropology* 152(3): 322–32.

De Witte, S., Hughes-Morey, G., Bekvalac, J. and Karsten, J. 2016. Wealth, health, and frailty in industrial-era London. *Annals of Human Biology* 43(3): 241–254.

Elders, J. 2015. Burial law and archaeology in the Church of England's jurisdiction. In R. Redmond-Cooper (ed.), *Heritage, Ancestry and Law: Principles, Policies and Practices in dealing with Historical Human Remains*. Great Britain, Institute of Art and Law.

Fiorato, V., Boylston, A. and Knusel, C. (eds.) 2007. *Blood Red Roses: The Archaeology of a Mass Grave from the Battle of Towton AD 1461*. Oxford, Oxbow Books.

Fowler, L. and Powers, N. 2012. *Doctors, Dissection and Resurrection Men: Excavations in the 19th-Century Burial Ground of the London Hospital, 2006*. Monograph 62. London, Museum of London Archaeology.

Gapert, R., Black, S. and Last, J. 2009a. Sex determination from the occipital condyle: discriminant function analysis in an eighteenth and nineteenth century British sample. *American Journal of Physical Anthropology* 138: 384–394.

Gapert, R., Black, S. and Last, J. 2009b. Sex determination from the foramen magnum: discriminant function analysis in an eighteenth and nineteenth century British sample. *International Journal of Legal Medicine* 123: 25–33.

Grainger, I., Hawkins, D., Cowal, L. and Mikulski, R. 2008. *The Black Death Cemetery, East Smithfield, London*. Monograph 43. London, Museum of London Archaeology.

Grainger, I. and Phillpotts, C. 2011. Excavations at the Abbey of St Mary Graces, East Smithfield, London. Monograph 44. London, Museum of London Archaeology.

Guidance on the Care of Human Remains in Museums, 2005. Department for Culture, Media and Sport. London.

Hassett, B. R. 2011a. Changing World, Changing Lives: Child Health and Enamel Hypoplasia in Post Medieval London. Unpublished PhD thesis, University College of London.

Hassett, B. 2011b.Technical note: estimating sex using cervical canine odontometrics: a test using a known sex sample. *American Journal of Physical Anthropology* 146(3): 486–489.

Henderson, M., Miles, A., Walker, D., Connell, B. and Wroe-Brown, Robon, 2013. 'He Being Dead yet Speaketh'. Excavations at Three Post-Medieval Burial Grounds in Tower Hamlets, East London, 2004-10. Monograph 64. London, Museum of London Archaeology.

Historic England 2005. Guidance for best practice for treatment of human remains excavated from Christian burial grounds in England. Available at: http://www.archaeologyuk.org/apabe/pdf/APABE_ToHREfCBG_FINAL_WEB.pdf

Hughes-Morey, G. M. 2012. Body Size and Mortality in Post Medieval England. Unpublished PhD thesis, University of Albany, USA.

Human Tissue Act 2004. Available at: http://www.legislation.gov.uk/ukpga/2004/30/contents

Inskip, S. A., Taylor, G. M., Zakrzewski, Z. R., Mays, S. A., Pike, A. W. G., Llewellyn, G., Williams, C. M., Lee, O. Y. C., Wu, H. H. T., Minnikin, D. E., Besra, G. S. and Stewart, G. R. 2015. Osteological, biomolecular and geochemical examination of an early Anglo-Saxon case of lepromatous leprosy. *PLOS ONE* 10(5): e0124282.

Ives, R. 2007. Metabolic Bone Disease and Cortical Bone Dynamics in Post-Medieval Urban Collections. Unpublished PhD thesis, University of Birmingham.

Ives, R., Mant, M., de la Cova, C. and Brickley, M. 2016. A large-scale palaeopathological study of hip fractures from post-medieval urban England. *International Journal of Osteoarchaeology* 27(2): 261–275.

Krakowka, K. 2015. Violence-related trauma from the Cistercian Abbey of St Mary Graces and a late Black Death cemetery. *International Journal of Osteoarchaeology* 27(1): 56–66.

Kendall, E. J., Montgomery, J., Evans, J. A., Stantis, C. and Mueller, V. 2013. Mobility, mortality, and the Middle Ages: identification of migrant individuals in a 14th century Black Death cemetery population. *American Journal of Physical Anthropology* 150: 210–222.

Lewis, M. E., Shapland, F. and Watts, R. 2016a. On the threshold of adulthood: a new approach for the use of maturation indicators to assess puberty in adolescents from medieval England. *American Journal of Human Biology* 28: 48–56.

Lewis, M. E., Shapland, F. and Watts, R. 2016b. The influence of chronic conditions and the environment on pubertal development. An example from medieval England. *International Journal of Paleopathology* 12: 1–10.

Lohman, J. and Goodnow, K. (eds) 2006. *Human Remains and Museum Practice.* New York and Oxford, Berghahn and Museum of London.

Magilton, J., Lee, F. and Boylston, A. (eds) 2008. *Lepers Outside the Gate: Excavations at the Cemetery of the Hospital of St James and St Mary Magdalene, Chichester, 1986-87 and 1993.* Chichester Excavations Volume 10, CBA Research Report 158. York, Council for British Archaeology.

Mant, M. and Roberts, C. A. 2015. Diet and dental caries in post-medieval London. *International Journal of Historical Archaeology* 19(1): 188–207.

Mant, M. 2016. 'Readmitted under urgent circumstances'. Uniting archives and the bioarchaeological record at the Royal London Hospital. In M. Mant and A. Holland (eds), *Beyond the Bones: Engaging with Disparate Datasets*: 37–60. London, Elsevier.

Mays, S., Harding, C. and Heighway, C. 2007. *Wharram XI: The Churchyard.* Wharram Settlement Series, Volume 11. Oxford, Oxbow Books.

Miles, A., Powers, N. and Wroe-Brown, R., and Walker, D. 2008. *St Marylebone Church and Burial Ground in the 18th to 19th Centuries: Excavations at St Marylebone School, 1992 and 2004-6.* Monograph 46. London, Museum of London Archaeology.

Miles, A. and White, W. and Tankard, D. 2008. *Burial at the Site of the Parish Church of St Benet Sherehog Before and After the Great Fire: Excavations at 1 Poultry, City of London.* Monograph Series 39. London, Museum of London Archaeology Service.

Molleson, T. and Cox, M., Waldron, H.A. and Whittaker, D. K. 1993. *The Spitalfields Project, vol. 2, The Anthropology: The Middling Sort.* CBA Research Report 86. York, Council for British Archaeology.

Müller, R., Roberts, C. A. and Brown, T. A. 2014. Biomolecular identification of ancient Mycobacterium tuberculosis complex DNA in human remains from Britain and continental Europe. *American Journal of Physical Anthropology* 153(2): 178–189.

Museum of London, 2006. *Policy for the Care of Human Remains in Museum of London Collections.* London, Museum of London.

Museum of London, 2011. *Policy for the Care of Human Remains in Museum of London Collections.* London, Museum of London.

Naveed, H., Abed, S. F., Davagnanam, I., Uddin, J. M. and Adds, P. J. 2012. Lessons from the past: cribra orbitalia, an orbital roof pathology. *Orbit* 31: 394–399.

Newman, S. 2015. *The Growth of a Nation: Child Health and Development in the Industrial Revolution in England, c. AD 1750-1850*. Unpublished PhD thesis, University of Durham.

Newman, S. L. and Gowland, R. L. 2016. Dedicated followers of fashion? Bioarchaeological perspectives on socio-economic status, inequality, and health in urban children from the industrial revolution (18th–19th C), England. *International Journal of Osteoarchaeology* 27(2): 217–229.

Nixon, T., McAdam, E., Tomber, R. and Swain, H. (eds) 2002. *A Research Framework for London Archaeology 2002*. London, Museum of London.

Plomp, K. A., Viðarsdóttir, U. S., Weston, D. A., Dobney, K. and Collard, M. 2015. The ancestral shape hypothesis: an evolutionary explanation for the occurrence of intervertebral disc herniation in humans. *BMC Evolutionary Biology* 15: 68.

Powers, N. (ed.) 2012. *Human Osteology Method Statement*. Museum of London Report. London, Museum of London.

Redfern, R. C. and Bekvalac, J. 2013. The Museum of London: an overview of policies and practice. In M. Giesen (ed.), *Caring for the Dead: Changing Attitudes Towards Curation of Human Remains in Great Britain*: 87-98. Woodbridge, Boydell & Brewer.

Redfern, R. C. and Bekvalac, J. 2014. The health of monastic orders in medieval London. In J. Clark, J. Hall., J. Keily, R. Sherris and R. Stephenson (eds), *'Hidden Histories and Records of Antiquity': Essays on Saxon and Medieval London for John Clark, Curator Emeritus, Museum of London*. London and Middlesex Archaeological Society Special Paper 17. London, London and Middlesex Archaeological Society.

Redfern, R. C., Gröcke, D. R., Millard, A. R., Ridgeway, V., Johnson, L. and Hefner, J. T. 2016. Going south of the river: a multidisciplinary analysis of ancestry, mobility and diet in a population from Roman Southwark, London. *Journal of Archaeological Science* 74: 11–22

Redfern, R., Marshall, M., Eaton, K. and Poinar, H. 2017. 'Written in bone': new discoveries about the lives and burials of four Roman Londoners. *Britannia*, 1-25. Doi: doi:10.1017/S0068113X17000216

Reeve, J. and Adams, M. 1993. *The Spitalfields Project, vol. 1, The Archaeology: Across the Styx*. CBA Research Report 85. York, Council for British Archaeology.

Roberts, C. and Cox, M. 2003. *Health and Disease in Britain*. Gloucestershire, Sutton Publishing Ltd.

Roberts, C. A. 2006. A view from afar. Bioarchaeology in Britain. In J. E. Buikstra and L. A. Beck (eds), *Bioarchaeology. The Contextual Analysis of Human Remains*: 417–439. Amsterdam and Boston, Elsevier.

Roberts, C. and Mays, S. 2011. Study and restudy of curated skeletal collections in bioarchaeology: a perspective on the UK and the implications for future curation of human remains. *International Journal of Osteoarchaeology* 21: 626–630.

Rodwell, W. 2011. *St Peter's, Barton-upon-Humber, Lincolnshire: Volume 1, History, Archaeology and Architecture*. Oxford, Oxbow Books.

Scheuer, L. 1998. Age at death and cause of death of the people buried in St Brides Church, Fleet Street, London. In M. Cox (ed.), *Grave Concerns - Death and Burial in England 1700-1850*: 100-109. CBA Research Report 113. York, Council for British Archaeology.

Schuenemann, V. J., Bos, K., DeWitte, S., Schmedes, S., Jamieson, J., Mittnik, A., Forrest, S., Coombes, B. K., Wood, J. W., Earn, D. J., White, W., Krause, J. and Poinar, H. N. 2011. Targeted enrichment of ancient pathogens yielding the pPCP1 plasmid of Yersinia pestis from victims of the Black Death. *Proceedings of the National Academy of Sciences* 108(38): E746–752.

Schuenemann, V. J., Singh, P., Mendum, T. A., Krause-Kyora, B., Jäger, G., Bos, K. I., Herbig, A., Economou, C., Benjak, A., Busso, P., Nebel, A., Boldsen, J. L., Kjellström, A., Wu, H., Stewart, G. R., Taylor, G. M., Bauer, P., Lee, O. Y., Wu, H. H., Minnikin, D. E., Besra, G. S., Tucker, K., Roffey, S., Sow, S. O., Cole, S. T., Nieselt, K. and Krause, J. 2013. Genome-wide comparison of medieval and modern Mycobacterium leprae. *Science* 341(6142): 179–183.

Shapland, F. and Lewis, M. E. 2013. Brief communication: a proposed osteological method for the estimation of pubertal stage in human skeletal remains. *American Journal of Physical Anthropology* 151 (2): 302–310.

Shaw, H., Montgomery, J., Redfern, R., Gowland, R. and Evans, J. 2016. Identifying migrants in Roman London using lead and strontium stable isotopes. *Journal of Archaeological Science* 66: 57–68.

Spencer, R. 2010. *Testing Hypotheses About Diffuse Idiopathic Skeletal Hyperostosis (DISH) Using Stable Isotopes and Other Methods.* Unpublished PhD thesis, University of Durham

Steel, F. L. D. 1960. Investigation of the skeletal remains of a known population. *Medicine and the Law* 1(1): 54–62.

Thomas, C. 2004. *Life and Death in London's East End: 2000 Years at Spitalfields.* London, Museum of London Archaeology.

Verlinden, P. 2015. *Child's Play? A New Methodology for the Identification of Trauma in Non-Adult Skeletal Remains.* Unpublished PhD Thesis, University of Reading.

Verlinden, P. and Lewis, M. E. 2015. Childhood trauma: methods for the identification of physeal fractures in nonadult skeletal remains. *American Journal Physical Anthropology* 157: 411–420.

Waldron, T. 2007. *St Peter's, Barton-upon-Humber, Lincolnshire. A Parish Church and its Community: Volume 2 The Human Remains.* Oxford, Oxbow.

Walter, B. S., DeWitte, S. and Redfern, R. C. 2016. Sex differentials of caries frequencies in medieval London. *Archives of Oral Biology* 63: 32–39.

Watts, R. 2013. Lumbar vertebral canal size in adults and children: observations from a skeletal sample from London, England. *HOMO Journal of Comparative Human Biology* 64: 120–128.

Werner, A. 1998. *London Bodies. The Changing Shape of Londoners from Prehistoric Times to the Present Day.* London, Museum of London.

Western, A. G. and Bekvalac, J. 2015. Digital radiography and historical contextualisation of the 19th century modified human skeletal remains from the Worcester Royal Infirmary, England. *International Journal of Palaeopathology* 10: 58–73.

Yaussy, S. L., DeWitte, S. N. and Redfern, R. R. 2016. Frailty and famine: patterns of mortality and physiological stress among victims of famine in medieval London. *American Journal of Physical Anthropology* 160(2): 272–83.

Chapter 3

The Grant Human Skeletal Collection and Other Contributions of J. C. B. Grant to Anatomy, Osteology, and Forensic Anthropology

John Albanese[1,2]

[1] Associate Professor, Department of Sociology, Anthropology and Criminology, University of Windsor, 401 Sunset Avenue, Windsor, Ontario, N9B 3P4, Canada

[2] Research Associate, Centre for Forensic Research, Simon Fraser University, 8888 University Dr., Burnaby, BC, V5A 1S6, Canada

Introduction

Dr. John Charles Boileau Grant is best known for his contributions to the instruction of anatomy. Grant was well known for his enthusiastic visual lectures on human anatomy, his strict discipline for staff and students, which fostered an atmosphere that challenged students to excel, and his encyclopedic knowledge of human anatomy (Breslin 1956; MacKenzie n.d.; Robinson 1988; Tobias 1992). Grant taught anatomy to thousands of medical students at the University of Manitoba, the University of Toronto and at the University of California at Los Angeles. Grant has also had a huge influence on the instruction of anatomy outside his own classroom: he was Chair of Anatomy first at the University of Manitoba and later at the University of Toronto; and the museum of anatomy that expanded under his direction beginning in 1930 is still an educational resource for medical students at the University of Toronto. He also authored three texts that were published in multiple editions: *A Method of Anatomy, Descriptive and Deductive* (1937); *Handbook for Dissectors* (with H. A. Cates, 1940); and *An Atlas of Anatomy* (1943). The *Handbook for Dissectors* now known as *Grant's Dissector* is in its 16th edition (Detton 2016) and *Grant's Atlas of Anatomy* has been translated into various languages including Italian, Japanese and Spanish, and is in its 14th edition (Agur and Dalley et al 2016).

Before World War II, almost all physical anthropologists were trained as anatomists, and Grant's anatomical teaching and research was very much intertwined with his anthropological interests. Grant's influence on osteology and forensic anthropology are less known. This chapter will provide a review of his contributions to these areas, describe the human skeletal reference collection that is named after him in order to illustrate its research potential, and provide some important historical context for interpreting the pattern of human variation that is sampled in the collection. While the focus in this chapter is on Grant's contributions to anthropology with a particular emphasis on osteology and forensic anthropology, it is impossible to separate the anatomy from the anthropology in the first half of the 20th century (Albanese 2003a, 2006; Hunt and Albanese 2005; Blakey 1987; Armelagos et al 1982).

The University of Edinburgh and the University of Durham: 1909–1919

Grant received his medical training at the University of Edinburgh and was exposed to the influence of several notable textbook authors, teachers and researchers. The University of Edinburgh has a long and distinguished tradition of anatomical and medical instruction, and also had an influence on the emerging field of physical anthropology (Tobias 1985). Grant studied anatomy under Daniel John Cunningham and likely used the first edition (published in 1902) or the second edition (published in 1905) of Cunningham's *Textbook of Anatomy*. As a student, Grant received several awards for his knowledge of anatomy and his skill at dissection, and upon graduating, was invited to work as a demonstrator for Cunningham from 1909 to 1911 (Tobias 1992). Grant's interest in physical anthropology may have originated during his time at Edinburgh where he had contact with Sir William Turner. Turner had collected a large number of identified skeletons at Edinburgh and had been an important influence on Robert J. Terry, who established the Terry Collection after his return to the United States (Hunt and Albanese 2005; Tobias 1985; Trotter 1981). Once he had the opportunity at the University of Manitoba and later at the University of Toronto, Grant also followed in the tradition of Turner in combining anatomy and anthropology (Tobias 1992).

From 1911 to 1913, Grant was a demonstrator of anatomy at what was then known as the University of Durham at Newcastle upon Tyne (now Newcastle University) under Professor R. Howden, who was the editor of *Gray's Anatomy* at the time. Grant is credited for assisting in the revision of the text and the preparation of several dissections for the illustrations in the 18th and 20th editions. During this time at Durham University, Grant also came in contact with one of his heroes and major influences, Sir Grafton Elliot Smith (Basmajian 1974). Smith also had an important influence on two other prominent anatomist-anthropologists: Raymond A. Dart and T. Wingate Todd (Hunt and Albanese 2005). Raymond A. Dart, who was later appointed Chair of Anatomy and established the Dart Collection at the University of Witwatersrand, was a senior demonstrator under Smith (Tobias 1985; Dart 1973). T. Wingate Todd, who was later appointed Chair of Anatomy and greatly expanded the Hamann-Todd Collection, was a lecturer under Smith and was responsible for processing and cataloguing the skeletons that Smith had acquired as part of the Nubian Archaeological Survey (Shapiro 1939).

On August 5, 1914, the day after Great Britain declared war on Germany, Grant wrote to the War Office to apply for a commission in the Royal Army Medical Corps, and he served as a medical officer from November 1914 to April 1919. After the war, Grant returned to his position as the demonstrator of anatomy at the University of Durham. Later that year, he was invited to apply for the Professorship and Chair of Anatomy at the University of Manitoba.

The University of Manitoba: 1919–1930

Grant arrived in Winnipeg to take up the Professorship and Chair of Anatomy at the University of Manitoba in October 1919, and held the position until 1930. Soon after his arrival, Grant began pursuing research in physical anthropology, which he described as his 'hobby' in an interview he gave decades later (Breslin 1956). In 1920, Grant served as the medical officer to an 'Indian Treaty Party' that traveled to York Factory and Churchill on Hudson Bay[1]. Fifty years later, he described how this trip affected him:

> 'On finding out that almost no work had been done on the Anthropometry (Physical Anthropology) of the North American Indians of Canada, it seems obvious that without further delay, data on the Indians should be collected before further intermixture with other races took place.' (Grant 1970).

With the guidance of Diamond Jenness, an early pioneer in anthropological research in Canada, Grant made several trips to various settlements in Manitoba and the Northwest Territories in Canada, to collect anthropometric data over three field seasons in 1927, 1928 and 1929. He published three volumes (Grant 1929, 1930, 1936) based on his data and data collected by Jenness in 1923. Grant modeled his research and writing after Louis R. Sullivan's (1920) paper 'Anthropometry of the Siouan Tribes' and Grant's publications followed an approach that was common for the period. As the quote above indicates, variation was considered within a racial paradigm and Grant collected a suite of standard measurements (stature, sitting height, etc.) and indices (cephalic, crural, etc.) and ABO blood group data. He published descriptive summary statistics and raw data for each individual in the sample. Grant also provided contextual information for his research sample and included brief sections on language affiliations, marriage practices and post-contact history (Grant 1929, 1930). Grant returned to the Northwest Territories in the summer of 1934 to collect additional anthropometric data but this research was never published (Anonymous 1934).

Soon after arriving in Winnipeg, Grant (1922) published his first paper on osteology, and was called upon to assist in forensic investigations where human skeletal remains were discovered. His work in this area could be considered the beginning of forensic anthropology in Canada. The earliest record of a request for assistance is dated to September 1921, and he continued to assist with these cases until the late 1950s (Breslin 1956). Unfortunately, details about specific cases are scarce because most of Grant's personal and research papers are missing. The author conducted an extensive search of the archival documents at the Anatomy Division (formerly Anatomy Department), the central archive, and the Department of Anthropology at the University of Toronto, as well as the medical archives at the University of Manitoba, and was able to locate only a few documents to piece together this history. It is likely that many of Grant's papers went missing when the Medical School at the University of Toronto moved to a new building about 10 years after Grant's retirement.

---

[1] The 'Numbered Treaties' were negotiated between various First Nations and the Government of Canada between 1871 and 1921. Treaty No. 5 (1875) was extended in 1910 to include this area of Manitoba on the shore of Hudson Bay. By the 1920s a treaty party would have included a government representative known as an 'Indian agent', officers of the Royal Canadian Mounted Police and a doctor. The political, social and economic implications of these treaties on First Nations were far reaching and persist today, but are beyond the scope of this chapter. Interested readers are referred to Beardy and Coutts' (1996) collection of oral histories of Cree Elders from York Factory; Coates and Morrison (1986) for a detailed report about Treaty No. 5; and Poelzer and Coates (2015) for more about treaties within a greater political and social context in Canada.

Grant began collecting individual bones or skeletal elements in an *ad hoc* fashion in the 1920s (for example, Basset 2015). He kept several of the archaeological crania that were sent to him and he also actively exchanged skeletal elements with M. R. Dreman at the Department of Anatomy, University of Cape Town. In the first half of the 20th century, it was a common practice by anatomists with an interest in skeletal variation to exchange skeletons with colleagues from around the world (Hunt and Albanese 2005). This typological approach to collecting skeletons, where only one or a few individuals of a specific group was considered a sufficient sample of variation, was common at the time (for the historical context of this problematic typological approach see Blakey 1987 and Armelagos et al 1982; and for the lasting impact of this approach on forensic anthropology see Albanese and Saunders 2006). The skeletal collection that Grant began almost a decade later was somewhat a departure from this typological approach.

The University of Toronto and the University of California at Los Angeles: 1930–1973

In 1930, Grant accepted an invitation to be the Chair of Anatomy at the University of Toronto, which he held until his retirement in 1956. Grant's first objective when he arrived in Toronto was to create:

> '*a teaching museum of anatomical material that would be* **used**... *It was designed that the specimens were placed in four-sided jars, set on revolving bases, hence each specimen had four surfaces to present, each was specially illustrated and labelled... the student, seated and with textbook or notes beside him, could study in comfort*' (Grant 1970, emphasis added).

This approach marks a major shift in how anatomy was taught, and over 75 years later, many of the original preparations on turntables are on display and are still used by students to study anatomy (Stewart n.d.).

Upon arriving in Toronto, Grant also began planning for what he referred to as the 'Anthropological Collection'. Like other anatomists at the time who were interested in physical anthropology, Grant set up a protocol whereby cadavers were processed for their skeletons after medical students had completed the dissections. Other notable anatomist-anthropologists whose collections are still readily available for research include Robert Terry at Washington University, St. Louis (the collection is now at the Smithsonian Institution, National Museum of Natural History); T. Wingate Todd at Western Reserve University in Cleveland (now Case Western Reserve; the collection is now at the Cleveland Museum of Natural History); George S. Huntington (Muller et al. 2017) at the College of Physicians and Surgeons in New York (now part of Columbia University; the collection is now at the Smithsonian Institution's National Museum of Natural History); and Raymond Dart at University of the Witwatersrand, Johannesburg, South Africa (Dayal *et al.* 2009; Hunt and Albanese 2005).

There are some important similarities that Grant shared with Todd and Terry. All three were charter members of the American Association of Physical Anthropologists (AAPA). Grant (1930) presented a paper at the first full meeting of the AAPA in April

1930, at the invitation of Aleš Hrdlička, the first physical anthropologist appointed at the Smithsonian Institution (Brown and Cartmill 2005; Blakey 1987). Also, like Todd, Grant was interested in comparative skeletal anatomy and also collected the skeletons of animals, though only a mounted skeleton of a male gorilla still hangs in Grant's Anatomy Museum. And like Terry, Grant went to great lengths to cross-check the documentary information for individuals in the collection. Both Terry and Grant operated under the assumption that information on death certificates (age at death and cause of death) that arrived with each cadaver in the anatomy department morgue should be considered suspect until the information was independently confirmed. The common practice for confirmation was to review the hospital records of the deceased or through correspondence with acquaintances of the deceased. However, each of these collectors had different research priorities and resources, and as a result, the respective collections vary considerably in size; in terms of what documentary data were collected, cross-checked and curated; and the condition of the skeletal remains.

Throughout the 1930s, both Todd and Terry retained the skeletons of almost all the cadavers that were dissected in their respective anatomy departments. Some of the skeletons were included in the collection while others were used for research and teaching. Todd personally collected a suite of anthropometric data from the first 2500 individuals in his collection; conducted groundbreaking research on age changes in the skeleton (for example, Todd 1920, 1921; Todd and Lyon 1924, 1925a, 1925b,1925c); published extensively on various osteological, anatomical and forensically relevant topics (see Krogman 1939 for a list of over 175 of Todd's publications grouped by subject area); and collected over 900 primate skeletons for comparative study (Hunt and Albanese 2005; Krogman 1939). Terry published only a few papers on osteology and anatomy (for example, Terry 1932). Instead, he focused on developing an ingenious method for collecting 'living' stature from cadavers (see Terry 1940); documenting the ante-maceration appearance of the cadavers (documents, plaster death masks, hair samples and scale photographs); and on cross-referencing and confirming the accuracy of documentary data for each individual in the collection, which he continued to work on after his retirement (Hunt and Albanese 2005). Collecting ceased or was considerably reduced when Terry and Todd vacated their respective Chairs of Anatomy. Todd died unexpectedly in 1938, and Terry retired in 1941. Mildred Trotter continued to add skeletons more selectively to the Terry Collection until 1965. She published extensively on anthropological and forensic topics including, but not surprisingly, landmark research on stature estimation from skeletal remains (for example, Trotter 1930, 1938, 1943; Trotter and Duggins 1948, 1950; Totter and Gleser 1951, 1952; see also Hunt and Albanese 2005; Conroy *et al* 1992). Interested readers should see Chapters 4 and 5 for more information on how these collections can still be used effectively for forensic and bioarchaeological research.

Grant's priorities were different than those of Todd and Terry, and as a result, there are some important differences in the Grant Collection. Grant only published one paper in the area of osteology (Grant 1922). However, with the assistance of Charlie Storton, Grant was able to divide his time and resources between the skeletal collection, the anatomy museum and his anatomy textbooks (Hall 2007). The skeletal collection and research in osteology were less of a priority for Grant as compared to Terry or Todd, and the Grant Collection, consisting of 202 skeletons, is significantly smaller than Terry's collection of over 1700 skeletons or Todd's collection of approximately 3300 skeletons. In addition, about half of the crania in the Terry Collection are intact and the rest typically have only one section, which was made to gain access to the brain for instruction in brain anatomy. In contrast, almost all the crania in the Grant Collection have both transverse and sagittal sections, which left each cranium in three or four parts. Only a few crania (approximately 5 individuals) are still intact because these cadavers were not suitably preserved for the anatomy courses and were never dissected.

The outbreak of World War II had a huge influence on Grant's work. At the age of 53, Grant tried to enlist in the armed forces but was rejected on medical grounds. He sought surgical treatment and again tried to enlist but was rejected because his contribution to the war effort in training doctors was considered more important than his service as a medical officer (Basmajian 1974; Robinson 1988). Before World War II, Germany was a major exporter of anatomy textbooks, and in the lead-up to the war, these textbooks became increasingly difficult to purchase outside of Germany. This lack of suitable textbooks was one of Grant's greatest motivations for beginning work on his textbooks (Storton, as quoted in Hall 2007). Grant served during the war to the best of his abilities by focusing his attention on his teaching, the anatomy museum and the publication of his anatomy texts.

During the war, little time and few resources were available for the skeletal collection. The retention of entire skeletons and specific skeletal elements was more selective and split among competing purposes from 1942 to 1945. Some skeletal elements with pathological conditions and 'ideal specimens' are not curated with their respective skeletons because these elements were used in anatomy demonstrations, to illustrate pathological conditions, and/or were prepared for display in the anatomy museum and to serve as models for the illustrations that appeared in Grant's publications. In a few cases, the skeletal elements were later returned to the skeletal collection to the respective individuals. In other cases, important elements of some individuals, usually with an interesting variation, are missing. For example, individual GR0394 is described as having 'left leg short and deformed' but the bones of the left leg are not in the collection. After 1945 and until the 1950s when collecting ceased, additions to the collection varied considerably from as high as 35 skeletons in 1946–47 to a low of just one in 1953–54 when the last skeleton was added to the collection.

Over 150 skeletons that were initially included in the collection were removed either because their ages could not be independently verified (approximately 100 individuals) or because those ages were overrepresented in the collection (approximately 50 individuals). Despite this correction, some ages are still overrepresented in the collection (see Figure 1), a common problem with skeletal collections derived from anatomical sources in the first half of the 20th century (Hunt and Albanese 2005). The documents associated with the individuals that were removed are still curated with the collection and have been a valuable source of information for understanding the current size and composition of the collection (see Watkins and Muller 2015 who describe a similar process for the Cobb Collection). It is not clear why two females with verified ages and three females whose ages were not verified were removed even though other females whose ages were not verified were kept in the collection.

After his retirement in 1956, Grant was appointed professor emeritus and curator of the Anatomy Museum. In 1961, he was invited to be a visiting professor of anatomy at the University of California at Los Angeles where he spent half of every year teaching anatomy until 1970. He was working on the 7th edition of his *Atlas of Anatomy* when he died in August 1973.

Grant's Legacy: James E. Anderson and Physical Anthropology in Canada

Although Grant's publications in the area of physical anthropology are limited, he made a significant contribution to the discipline through the training and supervision of James E. Anderson as an anatomist, medical doctor and physical anthropologist (Melbye 1995). Anderson graduated as an M.D. in 1953, and in 1956 was appointed as a lecturer in the Department of Anatomy at the University of Toronto, where he taught anatomy and a course in human osteology to pre-medical and anthropology students. In 1958, he was appointed as an assistant professor to the Department of Anthropology and worked closely with Lawrence Oschinsky to develop a series of graduate level courses in physical anthropology (for additional information on Oschinsky's contributions to physical anthropology in Canada, see Ossenberg 2001). In 1963, Anderson was appointed associate professor of physical anthropology at the State University of New York at Buffalo but returned to the University of Toronto in 1966, and brought several of his graduate students with him. In 1967, he was appointed chair of anatomy at the then recently established medical school at McMaster University in Hamilton, Ontario, where he also participated in the Burlington Growth Study in the nearby city of Burlington, Ontario. Although the Burlington Growth Study continues to be used primarily for craniofacial growth studies in dentistry and medical research (see Kulshrestha *et al* 2016; American Association of Orthodontists Foundation 2016), data from the Burlington Growth Study are still being used in various anthropological research projects, including several fairly recent doctoral dissertations by Clare McVeigh (1999) and Todd Garlie (2001) from McMaster University, and Sherry Fukuzawa (2002) from the University of Toronto. Anderson completed the 7th edition of *Grant's Atlas of Anatomy* after Grant's death, and then edited the 8th and 9th editions.

In addition to creating the first English-speaking graduate program in physical anthropology in Canada, Anderson's and Oschinsky's research also marked the beginning of a transformation in osteology from descriptive research to analytical research (Meiklejohn 1997; Melbye and Meiklejohn 1992). Oschinsky's influence was tragically limited by his unexpected death at the age of 45 in 1966. Anderson supervised the first two doctorates that were granted in physical anthropology in English-speaking Canada, which were awarded in 1969 to Michael Pietrusewsky and Jerry Melbye. Anderson continued to act as a supervisor to both students at the University of Toronto after he left for McMaster University (Melbye, in 2003, personal communication).

Anderson also directly supervised or influenced through his graduate courses an entire generation of osteologists in Canada and the United States, including Nancy Ossenberg, Queen's University, Kingston; Jerry Cybulski, Canadian Museum of History, Ottawa; Christopher Meiklejohn, University of Winnipeg; Michael Spence, Western University; Sonja Jerkic, Memorial University of Newfoundland; Jim MacDonald, Northeastern Illinois University; Joyce Siranni, SUNY-Buffalo; and Robert Sundick, Michigan State University, Kalamazoo (Jerkic 2001).

In turn, Jerry Melbye has had a huge influence on osteology and forensic anthropology in Canada and the United States through his casework and educational initiatives. Melbye was appointed to the Department of Anthropology at the University of Toronto after completing his doctorate and went on to a long career in research and education in skeletal biology and later in forensic anthropology. Between 2004 and 2009, Melbye was at the Texas State University at San Marcos where he helped to develop a new Ph.D. program in forensic anthropology. In Canada, he was instrumental in the establishment of Canada's first forensic science program at the University of Toronto, and he acted as a consultant when the forensic science program was established at the University of Windsor, where the author was appointed in 2004. He also taught thousands of undergraduate students, and supervised 30 doctoral and 54 master's students, who have distinguished themselves in their research and the training of another generation of skeletal biologists and forensic anthropologists. These include Susan Pfeiffer (1976), Shelley Saunders (1977), M. Anne Katzenberg (1983) and Christine White (1990), to name a few. Most professors of physical anthropology with interests in osteology in Canada were or are Anderson's students, Melbye's students, Melbye's students' students, or some combination of the preceding. The author of this chapter is a typical example: two of the author's biggest influences in osteology while an undergraduate at Western University (formerly the University of Western Ontario) were Christine White, who completed her Ph.D. with Melbye, and Michael Spence, who completed his master's degree with Anderson; Melbye supervised the author's master's degree at the University of Toronto, which included research involving the Grant Collection; Melbye's student, Shelley Saunders, was the author's doctoral supervisor at McMaster University (Hamilton, Canada); and the author was a Social Science and Humanities Research Council of Canada (SSHRC) postdoctoral fellow and worked with Susan Pfeiffer at the University of Toronto, conducting research that led to this publication.

Anderson's direct influence on physical anthropology continued well after he left the University of Toronto. For example, Katzenberg and Saunders co-edited several influential volumes, including *Biological Anthropology of the Human Skeleton* (the second edition of which was published in 2008). These thorough and comprehensive volumes were directly inspired by Anderson's publications. In a conversation with the author in 1998, Saunders explained how the goal for the first edition was to provide a more current volume to replace Anderson's guide for investigating past populations, *The Human Skeleton: A Manual for Archaeologists* (1962). See Jerkic (2001) for more information about Anderson's contributions to physical anthropology in Canada.

The Grant Human Skeletal Collection

Identified skeletal collections have been used extensively for medical and anthropological research for over 100 years (Hunt and Albanese 2006). The research value of anatomical and anthropological research of an identified skeletal collection is directly related to the quality of the documentary data associated with each individual and the collection as a whole (Hunt and Albanese 2006; Albanese 2003a; see also Chapters 4 and 5, this volume). Data quality and accuracy can only be assessed through a review of the protocol for collecting and the historical context of the collection period. This section will provide some additional context for the Grant Collection.

Every cadaver that arrived in the Anatomy Department at the University of Toronto was logged into the Anatomy Register and assigned a cadaver number in accordance with the Anatomy Act of Ontario (various dates; see Discussion section). This cadaver number is the same number associated with each skeleton currently in the collection. Until the late 1920s, cadaver numbers were assigned from one to 200. After the cadaver number 200 was reached, the numbering system started again at one. The cadaver number in conjunction with the year of death was used for identification purposes since there were never more than 200 cadavers processed in a given year. In the Anatomy Register that is still on file in the Division of Anatomy, there were five series of cadaver numbers (one to 200) before Grant arrived at the University of Toronto. In 1928, the sixth series of cadaver numbers was started. All of the individuals that are currently in the skeletal collection have Series Six cadaver numbers, with one exception (see below). After Grant arrived in Toronto and began making plans for the skeletal collection, cadaver numbers were not reset after the cadaver number 200 was assigned, but instead were sequentially assigned up to 1000. Series Seven numbers (beginning with number one again) were assigned to cadavers that arrived in the Anatomy Department starting in December 1954, but were not dissected until 1957, the year after Grant retired and several years after the last skeleton was added to the collection. The first skeletons were processed for the collection after the 1930–31 academic year. The last skeleton was added to the collection after the 1953 dissection course. From about 1931 to 1941, between 12 and 22 skeletons were added to the collection each year, which amounts to under 60% of the cadavers that were used for anatomical instruction.

The Grant Skeletal Collection, consisting of 202 skeletons in various degrees of completeness, and most of the documents associated with the collection were transferred to the Department of Anthropology at the University of Toronto in the mid-1980s. Partial cranial elements from an additional 80 individuals were also transferred at that time. These partial crania are numbered but do not have documents associated with them. Some of these skeletal elements pre-date Grant's arrival at the University of Toronto and were never part of the skeletal collection. Although the partial crania are curated with the full skeletons, they should be considered a separate collection. Using new data collected from documents located in the Department of Anatomy at the University of Toronto in 2003, the author has identified 41 of these individuals. However, age at death of these 41 individuals should be considered approximate. For the 39 unidentified partial crania with cadaver numbers below 200, the challenge is in determining whether they have a Series Five cadaver number or Series Six cadaver number.

Despite the fact that most of Grant's papers are not available in any archive, a series of original documents is still available for the skeletal collection. The documents that are curated with each skeleton varied in format over the 25 years of collecting but fall into two major types, which are referred to here as 'data forms' and 'assessment forms.' Data forms are full-page documents that are available for every individual (excluding the partial crania). The data forms varied over the decades of collecting, but they consistently include the following fields of data: name, age, age verified, sex, serial number (cadaver number), received from (source of the cadaver), date (of death), cause of death, date the dissection was begun, table number (where dissection took place) and a checklist of bones. Additional comments are either in the margin or in a designated notes/comments section on the form. Most of the data (age, age verified, sex, etc.) were consistently recorded on the forms, while other fields, such as the date of death, were often left blank.

Assessment forms are half page forms that are available for 33 of the 202 individuals in the collection and for over 150 individuals that are no longer in the collection with cadaver numbers from GR0437 to GR0837. Assessment forms have the following fields: cadaver number, name, date of death, place of death, patient number (if the person died in a hospital), cause of death, age at death and space to list the criteria to assess whether the age at death and cause of death should be considered correct. The assessment forms are invaluable for reconstructing the process by which cause of death and age at death data were verified, and why some individuals were kept in the collection while others were removed.

Besides sex (based on soft tissue), age at death and cause of death data should be considered accurate with very few exceptions. The author has reviewed all the documents available for the collection in the Department of Anthropology and the Division of Anatomy, including the documents associated with those individuals who were eventually removed from the collection. It is evident from these documents that Grant had a thorough and systematic procedure, similar to Terry's, to assess the accuracy of these two important fields of data. As noted above, death certificate information was not accepted as accurate without independent confirmation. The assessment forms illustrate the confirmation process, which followed a logical pattern where hospital records were reviewed, and individuals who knew the deceased were consulted. There is no evidence of bias in the confirmation of cause of death, and cause of death did not seem to be a criterion for inclusion or retention of an individual in the collection.

Although the assessment forms are available for only 33 of the individuals in the collection, there is evidence that the same process was followed for all the individuals who were included in the collection. In 35 cases (cadaver numbers less than GR0437) with no assessment form, the age at death listed in the Anatomy Register (described above) does not match the age at death that is designated as 'verified' on the data form. At first glance, this discrepancy suggests that the age data for these individuals may be suspect. In fact, the opposite is true. The age that was recorded in the Anatomy Register was listed on the death certificate and other documents that accompanied the cadaver to the anatomy department morgue. The Anatomy Register had to be kept in compliance with regulations in the Anatomy Act of Ontario and had to remain synchronized with the death certificate. The Anatomy Register should not be considered the definitive source for the correct age because it was never amended even after the correct age at death was determined through independent verification. These 35 'discrepancies' clearly illustrate that the systematic review process for cross-checking age at death data was applied to all the individuals (except the partial crania), and the age of death data were carefully verified even in those cases with no assessment forms.

The situation is different for the cause of death data. The cause of death information on data forms can be considered accurate but not always precise, regardless of whether an assessment form is available or not. For example, the cause of death for GR0640 is listed as 'bronchopneumonia' but on the Anatomy Register it is listed as 'left bronchopneumonia.' The greater detail available in the Anatomy Register is reliable, particularly when considered in conjunction with the source of the cadaver. Many of the individuals in the collection were transferred from hospitals or long term care facilities (i.e. sanatoriums) in the Greater Toronto Area, and thus, the cause of death would have been based on information collected over weeks or months of care and treatment immediately before death. It is not clear why there are small but potentially significant differences from the Anatomy Registry. However, pathological investigations involving individuals from the Grant Collection may be problematic if the information in the Anatomy Register is not reviewed. When considering all of the documents available in conjunction with the source of the cadaver, the cause of death data for the Grant Collection are more precise and accurate than death certificate data for this time period. However, it is important to note that the same caveats apply to the Grant Collection that would apply to any other cause of death data from the 1930s and 1940s. For example, 'senility' is described as the sole or contributing cause of death for two individuals. Because of regulations stemming from the Anatomy Act of Ontario, the Anatomy Register must be stored in fire-proof safe at the Division of Anatomy at the University of Toronto and it is not curated with the skeletal collection.

The protocol in place for the collection process was designed to avoid commingling of skeletal elements from different individuals. Among other things, the data form is a checklist of bones that followed the cadaver, starting with the dissection. There are also clear notes documenting if skeletal elements were removed (for example, 'humerus taken for museum'). Additionally, there are five documented cases where skeletal elements were mixed from several individuals at one time or another. In some cases, the mixing occurred when the cadavers were dissected. In other cases, the mixing occurred later in the collection process, but these elements were later removed. One example is presented to illustrate the thorough record keeping that was followed throughout the collection process, which has ensured the integrity of each individual skeleton in the collection. A comment in the remarks section of the data form for GR0345 states: 'on Feb 11, 1938 the right lower limb was found rotted with fracture of femur and poor injection and was replaced by left lower limb of subject no 170.' This intentional commingling was carefully documented so that it could be undone. When examining the skeleton of GR0345, it is clear that there is no evidence of mixing: the fractured right femur is present and is consistent in size and robusticity to the left femur, and there is no GR0170 in the collection. The limb from GR0170 was used only at the time of dissection for anatomical instruction and the correct leg was returned when the individual was added to the collection.

In a few other cases, there is evidence of accidental mixing of skeletal remains. These problems seem to be limited to a few ribs that are easily identified when the entire set of ribs is examined for these individuals. Significant problems with commingling occur in only one case: GR0185. The skeletal elements and the documentary data from two individuals are mixed. The problems stem from confusion over two individuals with the same cadaver number. One individual has a Series Five cadaver number and one individual has a Series Six cadaver number. This error occurred very early in the collection process because both the skeletal elements and the documentary information are a mix of Series Five-GR0185 and Series Six-GR0185. Series Five-GR0185 is a nearly complete skeleton of a 79-year-old female who died of arteriosclerosis in 1927. Series Six-GR0185 is represented by a partial cranium of a 60-year-old male that died of calculus pyonephrosis. The skeletal material from both of these individuals is easily separated. The information curated with the skeleton is a mix of data from both individuals: 60-year-old female with age verified with 'calculus pyonephrosis' as the cause of death. The male's age can be considered verified, but the female's age should be considered unverified because there is no evidence that the information for any the Series 5 individuals was rigorously verified.

One major difference that becomes obvious when comparing the Grant Collection to collections from the United States is the nature of the documentary data. Despite the racial view of human variation that was prevalent in physical anthropology in the first half of the 20th century (see Blakey 1987; Armelagos et al 1982), and evident in Grant's early anthropometric research, the documents associated with Grant's skeletal collection are remarkably lacking in racial designations. This lack of racialization is in very sharp contrast to other major research collections, such as the Hamann-Todd Collection and Terry Collection, and is more in line with the earlier Huntington Collection (now at the Smithsonian Institution) where the country of origin of recent immigrants was documented. These differences in documentation likely stem from the location where the collecting took place. In some cases, country of origin (for immigrants) and state of origin are documented for individuals in the Terry Collection. However, racial categories, particularly 'White' and 'Negro' (terms used in the original documents from the 1930s and 1940s), were used to designate individuals in the Terry and Haman-Todd collections to reflect the common popular views in St. Louis and Cleveland, two cities with significant African-American communities. Individuals in the collection were classified after death just as they were in life under the Jim Crow laws, decades before the equal rights movement in the United States. For example, standardized forms were used by Terry in St. Louis for documentary data for each individual in the collection. The form contains similar data as the Grant Collection data form described above with one significant difference. The first line of the form has spaces for series or year, cadaver number, name and skeleton number. These data had to be kept under the Anatomy Act in Missouri (various dates, see Discussion section). On line two, race was the first non-required datum that was recorded. In contrast, New York (Huntington Collection) and Toronto (Grant Collection) were, and still are, multiethnic and multicultural communities that served as major entry and settlement points for immigrants from various parts of the world (see also Pearlstein 2015). If and when they were categorized, individuals tended to be placed in arbitrary categories based on country of origin, rather than arbitrary categories based on perceived racial difference (see Blakey 1987 for more details).

Following the transfer of the Grant Collection to the Department of Anthropology, the collection was sent to the Cleveland Museum of Natural History (CMNH) for processing and cataloguing in late 1987, and was returned to the University of Toronto in 1988. At the time, the CMNH had recently completed processing the Hamann-Todd Collection and was one of the few institutions with the necessary expertise and facilities to treat skeletons in this condition. An electronic database of demographic data and skeleton inventory was created at the CMNH and is now curated with the skeletal material at the University of Toronto. The figures for this paper were generated using the CMNH database supplemented with data from the Anatomy Register, as well as information collected from the data forms and assessment forms that were not included in the CMNH database.

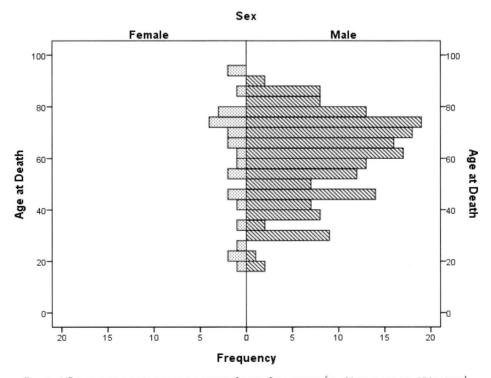

FIGURE 1. DEMOGRAPHIC COMPOSITION OF THE GRANT COLLECTION (N = 26 FEMALES; N = 176 MALES).

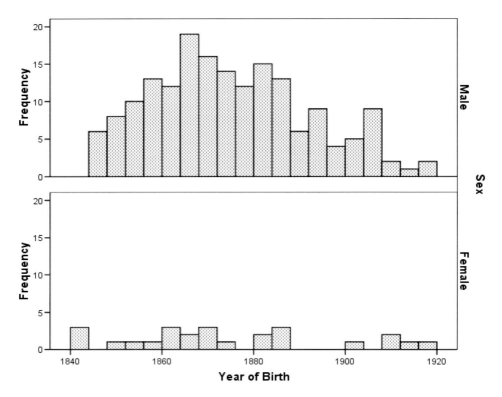

FIGURE 2. BIRTH COHORTS OF MALES AND FEMALES FROM THE GRANT COLLECTION. YEAR OF BIRTH WAS CALCULATED BY SUBTRACTING AGE FROM YEAR OF DEATH. BIRTH YEARS ARE GROUPED INTO 5 YEAR COHORTS (N = 26 FEMALES; N = 176 MALES).

Of the 202 individuals in the collection, 26 are female and 176 are male (Figure 1). Ages at death range from 17 to 93 years with the majority between 45 and 80 years (Figure 1). Ages at death were verified for all the males, and for 17 of the 26 females. However, considering the error described above for GR0185, age at death should be considered verified for only 16 females. Years of birth range from 1841 to 1918 with the majority falling between 1860 and 1890 (Figure 2). Year of birth was estimated by subtracting age at death from year of death. As previously mentioned, there are very few intact crania in the collection. Most crania have a transverse section that allowed for the removal of the calotte and a sagittal section of the cranium that divided the face into left and right halves. The sacrum and the mandible were typically sectioned sagittally, and in some cases, vertebral bodies have also been sectioned. Damage to other bones and the pattern of other missing elements in the rest of the skeleton is more sporadic. Despite the removal of some 'interesting' cases, there are many excellent examples of various pathological conditions, including perimortem fractures, poorly set but healed premortem fractures, and various lesions resulting from infectious diseases in a pre- and peri-antibiotic period. The most common causes of death are (in descending order) various types of cardiovascular disease (arteriosclerosis, myocarditis, etc), tuberculosis (all forms), various types of cancers, and bronchopneumonia (Table 3).

Cause of Death*	Frequency†	Percentage Frequency
Cardiovascular disease	72	36
Cancer	27	13
Tuberculosis	27	13
Bronchopneumonia	23	11

*Only those causes of death that are list more than 5 times are presented.
†This figure is the total number of times that a cause of death is listed as the sole cause of death or in conjunction with another cause of death.

TABLE 1. MOST COMMON CAUSES OF DEATH LISTED IN THE GRANT COLLECTION.

Discussion and Conclusion

The Grant Collection has been underutilized for a number of reasons. The size of the collection, the incompleteness of some skeletons, the very low number of females, and the condition of most of the crania have placed some limits on the collection's research potential. A second major impediment to research has been accessibility. Unlike the Terry Collection, which has been continually available for research, first at Washington University and now at the Smithsonian Institution, the Grant Collection followed a fate similar to the Hamann-Todd Collection. Both collections were not readily available for research for decades. In fact, both collections were nearly lost when there was no interest in their respective anatomy departments to maintain the collections. It was not until the 1980s that both collections were available for research after they were transferred to anthropology departments, processed, re-catalogued and stored in modern, easily accessible storage units.

Data from the Grant Skeletal Collection have been used in several master's and doctoral dissertations (for example, Sharman 2014; Sitchon 2003; Albanese 1997a), conference presentations (for example, Sharman 2004; Albanese 1997b, 1997c, 1997d, 1997e; Fairgrieve and Kaye 1995), and publications (for example, Albanese 2013; Albanese *et al.* 2008; Sitchon and Hoppa 2005; Usher 2002; Rogers 1999; Fairgrieve 1995; Bedford *et al.* 1993; Gruspier and Mullen 1991; Stuart-Macadam 1989; Lang 1987; Lovejoy *et al.* 1985). Much of the research has involved testing and developing forensic methods, particularly age and sex determination methods, and to a lesser extent, the collection has been used in paleopathological and paleoanthropological research. In general, despite some problems with all of the identified collections from this period (Komar and Grivas 2008; Ericksen 1982), research clearly shows that these collections in general are still invaluable for developing and testing modern forensic methods because of the quality of the documentary data (Albanese 2013; Albanese *et al.* 2008; Sharman 2014; see also Chapters 4 -7, this volume). The only limiting factor with the Grant Collection is the size of the collection, and it is best used in conjunction with other identified collections.

Skeletal collections such as the Terry, Todd, and Grant collections owe their existence to a unique set of conditions in the first half of the 20th century in the United States and Canada when most physical anthropologists were anatomists. Key individuals with research interests in physical anthropology were heads of their respective anatomy departments and could channel departmental resources to amassing and curating skeletal collections. Furthermore, they had legal access to cadavers for anatomical instruction (Terry 1940) through the anatomy acts in their respective jurisdictions at the state level in the Unites States and provincial level in Canada. The anatomy acts in many states and provinces were originally passed in the middle to late 19th century, and are directly derived from the Anatomy Act passed in England in 1832. The focus of that act was to allow the legal transfer of unclaimed cadavers to medical schools for the instruction of anatomy (Blake 1955). Thus, individuals who would have been buried 'at tax-payer expense' where transferred to qualified institutions. Richardson (2001) provides a comprehensive assessment of the social, political and economic context, as well as the ethical issues associated with anatomical instruction in England leading up to the passing of the act in 1832 (see Wilf 1989 for a similar discussion focused on New York).

The origins of the respective acts (Richardson 2001, Wilf 1989) and their application when the skeletal collections were amassed (Muller et al 2017) do raise some ethical concerns regarding anatomical instruction, skeletal collections and power relations within a society (as discussed in Chapter 9). However, the greater socio-economic and political context has continuously changed in the last 200 years in different jurisdictions, and a context-specific nuanced approach to ethical issues has some value and is essential to understand the patterns of variation in various collections (see Chapters 4 and 5, this volume). Although the acts have changed very little since the middle of the 19th century, in the Province of Ontario (Grant Collection), the State of Missouri (Terry Collection) and other jurisdictions, the source of cadavers gradually changed from exclusively unclaimed bodies at the beginning of the 20th century to almost exclusively donated bodies by the beginning of the 21st century. Yet, the differences between the skeletons of unclaimed individuals and donated bodies in the Terry Collection do not seem to be measureable (Ericksen 1982). Also, Charlie Storton, Grant's assistant for many years, noted in an interview in 2007, that while almost all of the individuals currently in the Grant Collection were unclaimed, the pathway for individuals was rather unique (as described in Hall 2007, see also Chapters 4 and 5, this volume).

There is little doubt that the greater power relations in society between dominant groups and marginalized groups are reflected in the demographic composition of the Terry Collection. It is not a coincidence or an accident that Black males make up the largest group and White females make up the smallest group in the Terry Collection (Hunt and Albanese 2005). These biases in the construction of collections have resulted in the use of the Terry and other collections to reinforce and perpetuate scientific and popular misconceptions of human variation (Albanese and Saunders 2006; see Chapter 4, this volume). However, the collections have been and continue to be invaluable for effectively critiquing racial and typological approaches to research and for providing alternatives to a racialized view of human variation that existed when the collections were amassed and that still persist. After Jesse Owens won multiple gold medals at the 1936 Olympics in Berlin, Cobb (1936) collected anthropometric data from Owens and other athletes and compared it to the detailed anthropometric data collected by Todd from cadavers that were to be included in the Hamann-Todd Collection. Cobb clearly demonstrated the non-concordance of so-called racial traits, and that it was training and not 'race' that resulted in multiple gold medals (see also Rankin-Hill and Blakey 1994; Chapter 9, this volume). More recently, in a series of papers the author and various colleagues have demonstrated using data collected from the Terry, Coimbra, Lisbon and Grant Collections that a racial approach is an impediment to developing effective methods that are applicable in forensic cases, and avoiding a racialized approach results in methods that are more accurate and easier to apply (Albanese *et al.* 2016a; Albanese *et al.* 2016b; Albanese 2013; Albanese *et al.* 2012; Albanese *et al.* 2008; Albanese and Saunders 2006; Albanese *et al.* 2005; Albanese 2003b).

Both physical anthropology and anatomy were very much shaped by the people, research paradigms and greater society in the first half of the 20th century. The development, during this critical period, of both disciplines has had lasting effects that are still seen today. The history of the modern emergence of physical anthropology and the development of the modern curriculum in anatomy were developed during this period. The skeletal collections are a material artifact of these distinct disciplines. While there are similarities between all of the skeletal collections from this period, there are also significant differences in the collections that reflect the research interests of the collector as discussed above, the state of the discipline, popular views of human variation in the greater society (in particular, 'race') and the interactions of these personal, discipline and societal biases (See Chapter 4, this volume). Only with historical context is it possible to understand how the collections were amassed in different parts of the world, which leads to a better understanding of human variation, and a better understanding of the history and current status of two major disciplines. Grant was an early pioneer in a much more hands-on approach to teaching anatomy, and the skeletal collection was an extension of that pedagogy. Grant's contributions to the instruction of anatomy, through his teaching, textbooks and anatomy museum, are enormous. Although his contributions to physical anthropology, osteology, forensic anthropology and bioarchaeology are less well known, they are still very significant. The Grant Collection is available for research to qualified individuals. Interested researchers should contact the Department of Anthropology at the University of Toronto for more information on how to gain access to the collection.

Acknowledgements

Thanks to Dr. Cynthia Kwok, Joan MacKenzie, Dr. Patricia Stewart and Dr. Jennifer Sharman. A special thanks to Pat Reed for facilitating access to the collection and documents in the Department of Anthropology and for a complete list of references involving data from the Grant Collection. This research was funded through a postdoctoral fellowship from the Social Science and Humanities Research Council of Canada. This paper is dedicated to Shelley Saunders, whose contributions to anthropological research cannot be expressed in words.

References Cited

Albanese, J. 1997a. *A Comparison of the Terry Collection and the Grant Collection Using the Head of the Femur and the Head of the Humerus: Implications for Determining Sex.* Master's thesis, University of Toronto.

Albanese, J. 1997b. A comparison of the Terry Collection and the Grant Collection using the head of the femur and head of the humerus: implications for determining sex. *Canadian Society of Forensic Science* 30: 167.

Albanese, J. 1997c. Similarities and differences between the Terry Collection and the Grant

Collection: the implications of collection and sample selection when developing sex determination methods. Paper presented at the *25th Annual Meeting of the Canadian Association for Physical Anthropology*, London, Canada, November 6–8.

Albanese, J. 1997d. Skeletal variability in recent North Americans: Implication for the development of new sex determination methods. Paper presented at the *25th Annual Meeting of the Canadian Association for Physical Anthropology*, London, Canada, November 6–8.

Albanese, J. 1997e. Changes in sexual dimorphism in North American 'whites': implications for the development of new forensic sexing methods. Paper presented at the *37th Northeastern Anthropological Association Meetings*, Montebello, USA, April 11–13.

Albanese, J. 2003a. *Identified Skeletal Reference Collections and the Study of Human Variation.* Unpublished PhD dissertation, McMaster University.

Albanese, J. 2003b. A Metric Method for Sex Determination Using the Hipbone and Femur. *Journal of Forensic Sciences* 48: 263–273.

Albanese, J. 2006. Contributions of J. C. B. Grant to Anthropology. Paper presented at the *34th Annual Meeting of the Canadian Association for Physical Anthropology*, Peterborough, Canada, October 25–28.

Albanese, J. 2013. A method for determining sex using the clavicle, humerus, radius and ulna. *Journal of Forensic Science* 58: 1413–1419.

Albanese J., Cardoso, H. F. V. and Saunders S. R. 2005. Universal methodology for developing univariate sample-specific sex determination methods: an example using the epicondylar breadth of the humerus. *Journal of Archaeological Science* 32: 143–152.

Albanese, J., Eklics, G. and Tuck, A. 2008. A metric method for sex determination using the proximal femur and fragmentary hipbone. *Journal of Forensic Science* 53: 1283–1288.

Albanese, J., Osley, S. E. and Tuck, A. 2012. Do century-specific equations provide better estimates of stature? A test of the 19th-20th century boundary for the stature estimation feature in Fordisc 3.0. *Forensic Science International* 219: 286–288.

Albanese J., Tuck, A., Gomes, J. and Cardoso, H. F. V. 2016a. An alternative approach for estimating stature from long bones that is not population- or group-specific. *Forensic Science International* 259: 59–68.

Albanese, J., Osley, S. E. and Tuck, A. 2016b. Do group-specific equations provide the best estimates of stature? *Forensic Science International* 261: 154–158.

Albanese, J. and Saunders, S. R. 2006. Is it possible to escape racial typology in forensic identification? In A. Schmitt, E. Cunha and J. Pinheiro (eds), *Forensic Anthropology and Medicine: Complementary Sciences From Recovery to Cause of Death*: 281–315., Totowa, NJ, Humana Press.

Agur, A. M. R. and Dalley, A. F. 2016. *Grant's Atlas of Anatomy*, 14th Edition. Alphen aan den Rijn (Netherlands), Wolters Kluwer.

American Association of Orthodontists Foundation (AAOF). 2016. *Craniofacial Growth Legacy Collection*, http://www.aaoflegacycollection.org/aaof_home.html.

Anderson, J. E. 1962. *The Human Skeleton: A Manual for Archaeologists*. Ottawa, National Museum of Canada.

Anonymous. 1934. *University of Toronto Monthly* 25 (1): 11.

Armelagos, G. J., Carlson, D. S. and Van Gerven, D. P. 1982. The theoretical foundations and development of skeletal biology. In F. Spencer (ed.), *A History of American Physical Anthropology 1930-1980*: 305–328. New York, Academic Press.

Basmajian, J. V. 1974. J. C. Boileau Grant. *Proceedings of the American Association of Anatomist* 1974: 176–178.

Basset, N. A. 2015. An osteobiographical account of the 'Red Indian' individual. Paper presented at the *43rd Annual Meeting of the Canadian Association for Physical Anthropology*, Winnipeg, Canada, October 28–31.

Bedford, M. E., Russell, K. F., Lovejoy, C. O., Meindl, R. S., Simpson, S. W. and Stuart-Macadam, P. I. 1993. Test of the multifactoral aging method using skeletons of known ages-at-death from the Grant Collection. *American Journal of Physical Anthropology* 91: 287–97.

Beardy, F. and Coutts R. 1996. *Voices from Hudson Bay: Cree Stories from York Factory*. Montreal, McGill-Queen's University Press.

Blake, John B. 1955. The development of American Anatomy Acts. *Journal of Medical Education* 30: 431–439.

Blakey, M. L. 1987. Skull doctors: intrinsic social and political bias in the history of American physical anthropology, with special reference to the work of Ales Hrdlicka. *Critique of Anthropology* 7: 7–35.

Breslin, C. 1956. J. C. Boileau Grant. *The Varsity* October 16: 8.

Brown, K. and Cartmill, M. 2005. 75 years of the annual AAPA meetings, 1930–2004. *American Journal of Physical Anthropology* Supplement 40: 79–80.

Coates, K. S. and Morrison, W. R. 1986. *Treaty Research Report: Treaty 5 (1875)*, Treaties and Historical Research Centre, Indian and Northern Affairs Canada. Available at : https://www.aadnc-aandc.gc.ca/eng/1100100028695/1100100028697#chp6.

Cobb, W. M. 1936. Race and runners. *Journal of Health and Physical Education* 7: 1–9.

Conroy, G., Phillips-Conroy, J., Peterson, R., Sussman, R. and Molnar, S. 1992. Obituary: Mildred Trotter, Ph.D. (February 2, 1899–August 23, 1991). *American Journal of Physical Anthropology* 87: 373–374.

Cunningham, D. J. 1902. *Cunningham's Textbook of Anatomy*. New York, William Wood and Company.

Cunningham, D. J. 1905. *Cunningham's Textbook of Anatomy, Second Edition*. New York, William Wood and Company.

Dart, R. A. 1973. Recollections of a reluctant anthropologist. *Human Evolution* 2: 417–27.

Dayal, M. R., Kegley, A. D. T., Strkalj, G., Bidmos, M. A. and Kuykendall, K. L. 2009. The history and composition of the Raymond A. Dart Collection of human skeletons at the University of the Witwatersrand, Johannesburg, South Africa. *American Journal of Physical Anthropology* 140: 324–35.

Detton, A. J. 2016. *Grant's Dissector, 16th Edition.* Alphen aan den Rijn (Netherlands), Wolters Kluwer.

Ericksen, M. F. 1982. How 'representative' is the Terry Collection? Evidence from the proximal femur. *American Journal of Physical Anthropology* 59: 345–50.

Fairgrieve, S. I. 1995. On a test of the multifactorial aging method by Bedford et al. (1993). *American Journal of Physical Anthropology* 97: 83–5.

Fairgrieve, S. I. and Kaye, B. H. 1995. Applications of fractal image analysis to forensic anthropology. Paper presented to the *47th Annual Meeting of the American Academy of Forensic Sciences*, Seattle, WA, February 13–18.

Fukuzawa, S. 2002. *A Longitudinal Examination of Heritability in the Developing Dental Arcade.* Unpublished PhD thesis, University of Toronto.

Garlie, T. 2001. *Stature, Mass, and Body Mass Index of Canadian Children.* Unpublished PhD thesis, McMaster University.

Grant, J. C. B. 1922. Some notes on an Eskimo skeleton. *American Journal of Physical Anthropology* 5: 267–71.

Grant, J. C. B. 1929. *Anthropometry of the Cree and Saulteaux Indians in Northeastern Manitoba.* Ottawa, F. A. Acland, King's Printer.

Grant, J. C. B. 1930. *Anthropometry of the Chipewyan and Cree Indians of the Neighbourhood of Lake Athabaska.* Ottawa, National Museum of Canada.

Grant, J. C. B. 1936. *Anthropometry of the Beaver, Sekani and Carrier Indians.* Ottawa, Canada Department of Mines, National Museum of Canada.

Grant, J. C. B. 1937. *A Method of Anatomy: Descriptive and Deductive.* Baltimore, William Wood and Co.

Grant, J. C. B. and Cates, H. A. 1940. *A Handbook for Dissectors.* H. A. Cates Baltimore, Wilkins.

Grant, J. C. B. 1943. An Atlas of Anatomy. Baltimore, Williams and Wilkins.

Grant, J. C. B. 1970. *Vitae.* Archival document curated in the Anatomy Division, University of Toronto.

Gruspier, K. L. and Mullen, G. J. 1991. Maxillary suture obliteration: a test of the Mann method. *Journal of Forensic Science* 36: 512–519.

Hall, J. 2007. A macabre collection. *Toronto Star* July 7. Available at : https://www.thestar.com/news/2007/07/07/a_macabre_collection.html

Hunt, D. R. and Albanese, J. 2005. History and demographic composition of the Robert J. Terry anatomical collection. *American Journal of Physical Anthropology* 127: 406–17.

Jerkic, S. M. 2001. The influence of James E. Anderson on Canadian Physical Anthropology. Proceedings of the symposium *Out of the Past: The History of Human Osteology at the University of Toronto*, University of Toronto, October 25, 2000.

Katzenberg, M. A. and Saunders, S. R. (eds). 2008. *Biological Anthropology of the Human Skeleton.* New Jersey, John Wiley and Sons.

Komar, D. A. and Grivas, C. 2008. Manufactured populations: what do contemporary reference skeletal collections represent? A comparative study using the Maxwell Museum documented collection. *American Journal of Physical Anthropology* 37: 224–33.

Krogman, W. M. 1939. Contributions of T. Wingate Todd to anatomy and physical anthropology. *American Journal of Physical Anthropology* 25: 145–86.

Kulshrestha, R., Trivedi, H., Tandon, R., Singh, K., Chandra, P., Gupta, A. and Ahmad, I. 2016. Growth and growth studies in orthodontics – a review. *Journal of Dentistry and Oral Care* 2: 1– 5.

Lang, C. 1987. Osteometric differentiation of male and female hip bones: an exploratory analysis of some unorthodox measurements. *Canadian Review of Physical Anthropology* 6: 1–9.

Lovejoy, C. O., Meindl, R. S., Pryzbeck, T. R. and Mensforth, R. P. 1985. Chronological metamorphosis of the auricular surface of the ilium: a new method for the determination of adult skeletal age at death. *American Journal of Physical Anthropology* 68: 15–28.

MacKenzie, R. n.d. History of the Anatomy Department at the University of Toronto. Unpublished manuscript written in 1973-74.

McVeigh, C. 1999. *Variability in Human Tooth Formation: A Comparison of Four Groups of Close Biological Affinity (England, Canada).* Unpublished PhD thesis, McMaster University.

Meiklejohn, C. 1997. Canada. In F. Spencer (ed.), *History of Physical Anthropology: An Encyclopedia*: 245–249. New York, Garland Publishing.

Melbye, J. 1995. Dr. James E. Anderson, 1926–1995: an obituary. *The Connective Tissue* 11: 11.

Melbye, J. and Meiklejohn, C. 1992. A history of physical anthropology and the development of evolutionary thought in Canada. *Human Evolution* 7: 49–55.

Muller, J. L., Pearlstein, K. E. and de la Cova, C. 2017. Dissection and documented skeletal collections: embodiments of legalized inequality. In K. C. Nystrom (ed.), *The Bioarchaeology of Dissection and Autopsy in the United States*: 185–201. Cham, Springer International Publishing.

Ossenberg, N. S. 2001. Lawrence Oschinsky: the contribution to Canadian osteology of a classical anthropologist. Proceedings of the symposium *Out of the Past: The History of Human Osteology at the University of Toronto*, held at the University of Toronto, October 25, 2000.

Pearlstein, K. E. 2015. Health and the Huddled Masses: An Analysis of Immigrant and Euro-American Skeletal Health in 19th Century New York City. Unpublished PhD thesis, American University.

Poelzer, G. and Coates, K. S. 2015. *From Treaty Peoples to Treaty Nation: A Road Map for All Canadians*. Vancouver, UBC Press.

Rankin-Hill, L. M. and Blakey, M. L. 1994. W. Montague Cobb (1904-1990): physical anthropologist, anatomist, and activist. *American Anthropologist* 96: 74–96.

Richardson, R. 2001. *Death, Dissection, and the Destitute*. New York, Routledge & Kegan Paul.

Robinson, C. L. N. 1988. Further remembrances of that revered anatomist, Dr. J. C. B. Boileau Grant. *Canadian Journal of Surgery* 31: 203–204.

Rogers, T. L. 1999. A visual method of determining the sex of skeletal remains using the distal humerus. *Journal of Forensic Science* 44: 57–60.

Rotter, M. 1930. The form, size, and color of head hair in American whites. *American Journal of Physical Anthropology* 14: 433–445.

Rotter, M. 1938. A review of the classification of hair. *American Journal of Physical Anthropology* 24: 105–126.

Shapiro, H. L. 1939. Thomas Wingate Todd. *American Anthropologist* 41: 458–64.

Sharman, J. A. 2004. Sex determination using the clavicle: The Grant Collection. Paper presented at the *Canadian Association for Physical Anthropology Meetings*, London, Ontario.

Sharman, J. A. 2014. *Age, Sex and the Life Course: Population Variability in Human Ageing and Implications for Bioarchaeology.* Unpublished PhD thesis, Durham University.

Sitchon, M. L. 2003. *Estimation of Age from the Pubic Symphysis: Digital Imaging Versus Traditional Observation.* Unpublished Master's thesis, University of Manitoba.

Sitchon, M. L. and Hoppa, R. D. 2005. Assessing age-related morphology of the pubic symphysis from digital images versus direct observation. *Journal of Forensic Science* 50: 791–795.

Stewart, P. n.d. *History of the Department of Surgery.* Toronto, University of Toronto.

Stuart-Macadam, P. I. 1989. Grant collection available for study. *Palaeopathology Newsletter* 67.

Sullivan, L. R. 1920. Anthropometry of the Siouan Tribes. *Proceedings of the National Academy of Sciences of the United States of America* 6: 131–134.

Terry, R. J. 1932. The clavicle of the American Negro. *American Journal of Physical Anthropology* 3: 351–379.

Terry, R. J. 1940. On measuring and photographing the cadaver. *American Journal of Physical Anthropology* 26: 433–447.

Tobias, P. V. 1985. History of physical anthropology in Southern Africa. *Yearbook of Physical Anthropology* 28: 1–52.

Tobias, P. V. 1992. J. C. Boileau Grant and the changing face of anatomy. *Clinical Anatomy* 5: 409–416.

Todd, T. W. 1920. Age changes in the pubic bone: I. The male white pubis. *American Journal of Physical Anthropology* 3: 285–334.

Todd, T. W. 1921. Age changes in the pubic bone: II, the pubis of the male negro-white hybrid; III the pubis of the white female; IV the pubis of the female negro-white hybrid. *American Journal of Physical Anthropology* 4: 1–70.

Todd, T. W. and Lyon, D. W. 1924. I Endocranial suture closure in the adult males of white stock. *American Journal of Physical Anthropology* 7: 325–384.

Todd, T. W. and Lyon, D. W. 1925a. II Ectocranial suture closure in the adult males of white stock. *American Journal of Physical Anthropology* 8: 23–45.

Todd, T. W. and Lyon, D. W. 1925b. Endocranial suture closure in the adult males of negro stock. *American Journal of Physical Anthropology* 8: 47–71.

Todd, T. W. and Lyon, D. W. 1925c. Ectocranial suture closure in the adult males of negro stock. *American Journal of Physical Anthropology* 8: 149–168.

Trotter, M. 1943. Hair from Paracas Indian mummies. *American Journal of Physical Anthropology* 1: 69–75.

Trotter, M. 1981. Robert James Terry, 1871–1966. *American Journal of Physical Anthropology* 56: 503–508.

Trotter, M. and Duggins, O. H. 1948. Age changes in head hair from birth to maturity. I. Index and size of hair of children. *American Journal of Physical Anthropology* 6: 489–501.

Trotter, M. and Duggins, O. H. 1950. Cuticular scale counts of hair of children. *American Journal of Physical Anthropology* 8: 467–484.

Trotter, M. and Gleser, G. C. 1951. Trends in stature of American whites and negroes born between 1840 and 1924. *American Journal of Physical Anthropology* 9: 427–440.

Trotter, M. and Gleser, G. C. 1952. Estimation of stature from long bones of American whites and negroes. *American Journal of Physical Anthropology* 10: 463–514.

Usher, B. M. 2002. Reference samples: the first step in linking biology and age in the human skeleton. In R. D. Hoppa and J. W. Vaupel (eds.), *Paleodemography: Age Distributions from Skeletal Samples*: 29–47. Cambridge, Cambridge University Press.

Watkins, R. and Muller, J. 2015. Repositioning the Cobb Human Archive: the merger of a skeletal collection with its texts. *American Journal of Human Biology* 27: 41–50.

Wilf, S. R. 1989. Anatomy and punishment in late eighteenth-century New York. *Journal of Social History* 22: 507–30.

Chapter 4

Strategies for Dealing with Bias in Identified Reference Collections and Implications for Research in the 21st Century

John Albanese[1,2]

[1] Associate Professor, Department of Sociology, Anthropology and Criminology, University of Windsor, 401 Sunset Avenue, Windsor, Ontario, N9B 3P4, Canada

[2] Research Associate, Centre for Forensic Research, Simon Fraser University, 8888 University Dr., Burnaby, BC, V5A 1S6, Canada

Introduction

Identified skeletal reference collections can be used to investigate a very broad range of research questions that can only be effectively addressed with a combination of skeletal and documentary data. Most methods used by forensic anthropologists for preliminary identification are only possible because of this unique combination of skeletal and documentary data. These methods include a range of approaches for estimating sex, age at death, stature and in some jurisdictions race or ancestry.

Because many identified reference collections were amassed around the middle of the 20th century or earlier, these collections have been described as biased (Komar and Grivas 2008) and not useful for developing methods that are applicable to 21st century forensic cases. There has been a trend to begin new identified skeletal collections and virtual collections or databases of skeletal data at various institutions (for example, Ferreira *et al.* 2014; Marinho and Cardoso 2013; Ousley and Jantz 1998; see Chapter 7, this volume). Certainly, more collections mean more potential research, but the problem is not with the old collections, but rather with the old way of looking at identified reference collections. Newer or more recent collections are different but not necessarily better sources of data, and if they correctly identify and address any issues of bias in the older collection, the newer collections introduce their own biases. None of these collections is a random sampling of the biocultural, geographic or statistical population they were drawn from or *Homo sapiens* as a species, and the impact of the bias varies considerably with the research question being addressed. All collections are biased *samples* and not biological or statistical *populations*, and a random sample of a biased sample can only result in a biased sub-sample being used for research. Identifying the biases in any collection is the critical first step to selecting the best sample possible to address specific research questions.

The term bias tends to have a negative connotation, but if bias is defined as directional rather than random error, it can and does have a huge unintended or unidentified influence on research and can produce very misleading results. There are two major sources of bias in the reference collections: 1) the *source* of the skeletons and documentary data, and 2) the collection *process* which can magnify some biases, address others and create new ones. Furthermore, the value of these collections is directly due to the fact that documentary information accompanies each skeleton and the collection as a whole. The collection process has not only had an influence on which skeletons were/are included in the collection. The collection process has had an enormous impact on what documentary data were retained, independently verified, and readily curated with the skeletons. This collection protocol, and even after collecting has stopped, the curatorial process are not static. The source and nature of the biases vary over time and are highly influenced by a number of factors including, first, the research interests of the collectors/curators; second, the discipline parameters for research; and third, popular views held by the greater society in which the collecting occurs.

The main goals of this chapter are to present a theoretical model for assessing the level of bias and its impact on research; and to demonstrate how various biases inherent in *all* collection can be identified, assessed and controlled for, or even exploit to maximize the research potential of the collection to address specific research questions. The model has proven to be a good tool for sampling reference collections, interpreting variation in those samples and our species, and to develop and test methods that are applicable in forensic and archaeological contexts (Albanese 2003a, 2010, 2013; Albanese and Saunders 2006; Albanese *et al.* 2005, 2008, 2012, 2016a, 2016b; Cardoso *et al.* 2016). Various examples used to illustrate the model will be drawn from the anatomical-derived Terry Collection (Hunt and Albanese 2005; Terry 1940) and the cemetery-derived Coimbra Collection[1] (Rocha 1995; Santos 2000). However, examples from the Hamann-Todd Collection (Hunt and Albanese 2005; Meindl *et al.* 1990), Lisbon Collection (Cardoso 2006), Grant Collection (see Chapter 3, this volume), the St Thomas' database (Saunders *et al.* 1995) and the Forensic Anthropology Database or FDB (Ousley and Jantz 1998) will also be discussed.[2] Readers are encouraged to also see the Chapter 5 of this volume for additional complementary information with a focus on age at death data.

A Model for Identifying and Assessing the Sources and Nature of Bias: An Approach to Pursuing Research Using Identified Skeletal Collections

Identified *reference* collections should be considered a subset of identified collections because they were amassed for general research purposes and for anatomical instruction, and not necessarily to reconstruct the population from which they were derived. Many of the most widely used collections were derived from anatomical sources where the skeletons were retained after the cadaver was used for the

---

[1] In this chapter, 'Coimbra Collection' refers to the series of 505 identified skeletons, known as the *Colecção de Esqueletos Identificados,* which consists of individuals who died between 1904 and 1936 and who were exhumed from the *Cemitério Municipal da Conchada* in Coimbra, Portugal.

[2] It is impossible to provide the detailed context for the sources of bias for each collection. Selective examples from a breadth of collections are intended to illustrate the potential of the theoretical model for assessing the nature and magnitude of bias in all collections and its possible impact on research. Where possible, the reader is referred to various sources for additional information.

anatomical training of doctors and dentists (Hunt and Albanese 2005; see also Chapter 3, this volume). However, in some cases, reference collections were derived from cemetery sources (Rocha 1995). All reference collections have been constructed (or are being constructed) by people working within historic, economic and discipline-specific contexts. Within these contexts, choices were made regarding whose skeleton was included in a collection, and what demographic data were verified and readily curated. Furthermore, when the period for collecting was long or occurred over several phases, the criteria for inclusion may have varied over time due to obvious reasons such as funding and resources and/or due to a number of other factors such as research interests and major trends in the discipline. Figure 1 is a variant of a graphical representation of a model that was first presented by Albanese (2003b). It represents a multi-disciplinary approach to identify bias in reference collections, and to work around or exploit the biases to address specific research questions. The model was developed through the combination of cemetery studies theory (Hoppa 1996, 1999; Saunders and Herring 1995a) and the New Biocultural Synthesis (Goodman and Leatherman 1998a) modified for research involving reference collections. The ultimate goal is to maximize the research potential of the collections and not to discredit the 'older' collections.

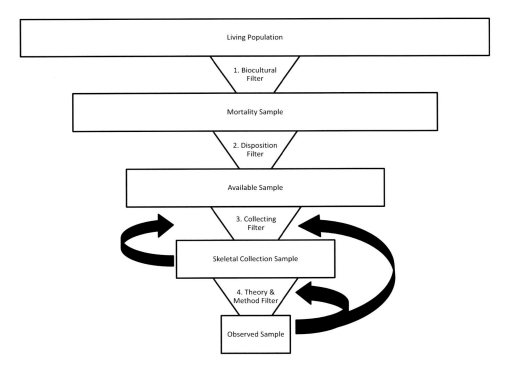

FIGURE 1. A GRAPHICAL REPRESENTATION OF A MULTI-DISCIPLINARY APPROACH TO IDENTIFY BIAS IN REFERENCE COLLECTIONS, AND TO WORK AROUND OR EXPLOIT THE BIASES TO ADDRESS SPECIFIC RESEARCH QUESTIONS (VARIANT OF ALBANESE 2003B AND BASED, IN PART, ON HOPPA 1996, 1999; SAUNDERS AND HERRING 1995A; GOODMAN AND LEATHERMAN 1998A).

Cemetery Studies Theory and Methods

The excavation of various types of cemeteries associated with record-keeping institutions such as churches, hospitals, workhouses, etc. from the historic period in Europe and North America have provided opportunities to test and evaluate the reconstruction of past populations through the comparison of skeletal data with the available documentary data for those populations (Saunders and Herring 1995b). The edited volume *Grave Reflections* is a collection of this research and a synthesis of key concepts that are relevant to the study of mortuary samples, i.e. all the individuals who were included in a cemetery, crypt or graveyard. Various examples described in that volume illustrate that representativeness of skeletal samples can be assessed at several levels, and that with large samples in good condition it is possible to make substantiated statements about past populations (Saunders and Herring 1995b). However, the editors of the volume note that,

> '...skeletal samples often reflect the uniqueness of the sample of individuals who were interred in a site, sometimes under very unusual circumstances. In many instances, it is unlikely that mortality samples will accurately represent broader population groups or communities simply because there are probably a multiplicity of communities that existed in the past whose composition fluctuated over time. This is certainly true of many historic period groups who lived when population size had become larger and more complex. Equally sobering is the distinct possibility that certain biological characteristics underwent secular change, which makes the deciphering of features, such as skeletal age changes, particularly difficult.' (Saunders and Herring 1995b:2).

If the phrase 'interred in a site' is replaced with 'included in a reference collection,' they are describing many of the issues associated with reference collections including anatomical-derived collections such as the Terry Collection or the cemetery-derived collections such as the Coimbra Collection. Considering reference collections in this theoretical framework previously applied to archaeological collections begins with accepting that all reference collections are biased. However, through the use of complementary historical documents, it is possible to assess the nature and impact of bias, and to overcome many problems with the analysis of these samples (Saunders *et al.* 1995).

In *Grave Reflections*, there are multiple examples[3] that illustrate approaches that can be used to assess the nature and source of the bias if skeletal and documentary data are considered. A major source of bias in any cemetery or reference collection is filtering of the sample at a number of different levels in space and time which influences the size and nature of the analyzable sample. This bias is clearly illustrated by Molleson

---

[3] Two examples are briefly presented here to illustrate a general model for considering bias and how it may apply to reference collections, and not an exhaustive description of the detailed context for the sources of bias for each collection. See Saunders and Herring (1995a) for additional examples; Scheuer and Bowman (1995) for more information about St. Bride's; Molleson (1995) and Reeve and Adams (1993) for details about Spitalfields; Tarlow 2015 for a compilation on burials in Europe during this period; Chapter 2, this volume, for a description of identified collections at the Museum of London.

using an area graph of documentary and skeletal data from Christ Church Spitalfields, London cemetery collection (1995:202, Figure 2). Of the 68,000 individuals buried at Christ Church, only 968 burials were excavated in the crypt of the church. Of these 968, only 389 were both positively identified through coffin plates and preserved well enough to be analyzed. In another example, Scheuer and Bowman (1995) discuss some factors that have affected the size and composition of an analyzable crypt sample using data from St. Bride's Church crypt, London, but which are applicable to a number of different specialized burials. Included in these factors are the level of preservation of skeletal material and coffin plates (i.e. the source of the demographic data), excavation procedures (a problem for St. Bride's Church crypt, see Chapter 2), and socio-economic factors that resulted in burials in special locations (i.e. the crypt) versus other areas. The nature and magnitude of the filters will be different for all collections, but the end result is the same. Bias is introduced or magnified in the study sample through the filtering. The top part of Figure 1 which borrows heavily from Hoppa (1996:52, Figure 3.1; see also Hoppa 1999) is a general model of this process.

Biased collections do not necessarily produce biased results. The assessment of the bias is the critical step in determining the level of representativeness of the sample (Saunders and Herring 1995b). Since in some ways they are highly specialized burial samples, reference collections can be considered in the same theoretical framework as archaeological mortuary collections, and thus, levels of representativeness in the reference collections and the effects of the bias will vary depending on the research questions being addressed. Once the bias is identified, it is possible to assess the level of representativeness of the collection, and to sample the skeletal collection in order to minimize or control the effects of the bias. Furthermore, it is also possible to sample the collection so that the bias may actually be exploited to maximize the research potential of the study sample to address specific research questions. In one example of this approach, Cunha (1995) used the Coimbra Collection to investigation skeletal stress indicators. Several biases become obvious after a preliminary review of the documentary resources that are available for the Coimbra Collection (Cunha 1995): first, the collection is a sample of individuals from lower socio-economic classes; and second, a relatively restricted geopolitical area is sampled in the collection. By sampling only those adults from the collection that were actually born in the District of Coimbra in order to control for social, economic and political differences between districts, Cunha took full advantage of the documentary information and further exploited the biases to create a near-perfect situation for addressing a specific research question (for a similar example using the data from both the Coimbra and Lisbon Collections see Alves Cardoso *et al.* 2016).

The New Biocultural Synthesis

Goodman and Leatherman (1998b:19-20) describe five interrelated concepts of the New Biocultural Synthesis (NBS): 1) a rejection of reductionist indicators and consideration of the effects of power relations on forming local environments; 2) the importance of the links between the local and the global, and the interaction of the two; 3) local and global historical context are critical to understanding the direction of social change, and the biological consequences of change including evolutionary change; 4) the environment only takes on meaning in relationship to the subject, and thus, humans shape, and at the same time, are shaped by their environment; 5) the ideology, knowledge, and perspectives of subjects and scientists are essential to understand human action. The NBS approach provides additional explanatory power for unraveling and interpreting the sources of variation in mortuary samples. However, when the focus of research involves reference collections and the goal is not to reconstruct the population from which the collection was derived, the third issue with an emphasis on historical context, and the fifth issue with an emphasis on the ideology, knowledge, and perspectives of the researchers are particularly significant and fill gaps that cannot be addressed in the cemetery studies model alone.

With the exception of problems associated with excavation methods, the filters that influence the analyzable mortuary samples are imposed by the cultures and societies of the deceased and not by the researchers studying the samples (Hoppa 1996, 1999). With most research involving past populations, the researcher is usually temporally, and often culturally distant from the deceased. Socio-economic and cultural criteria such as religion, age, wealth, status, occupation and views of gender of the deceased at the time of death resulted in inclusion in a 'typical' or a 'special' burial (Hoppa 1996, 1999). Similar criteria have had an effect on which individuals are included in a reference collection; however, the researcher-collectors (Drs Todd, Terry, Trotter, Grant, etc.) working in the context of their research discipline and society, which initially is also the society of the deceased, have played an important role in determining which individuals were included and retained in the reference collections. The ideology, knowledge and perspective of past researcher-collectors and current researchers have had and continue to have direct effects on theoretical and methodological issues including (1) construction of the demographic profile of the reference collections; (2) what documentary data were systematically collected, catalogued, and made readily available for research; (3) approaches to sampling the collections; and (4) research questions that have been investigated using the collections. Thus, the fifth issue (researcher bias) described by Goodman and Leatherman (1998b), particularly in conjunction with the third issue (historical perspective) may have the greatest effect, but seems to be ignored in most research involving reference collections.

From Living Population to Available Sample

Beginning at the top in Figure 1, the first three levels (living population, mortality sample, available sample) and the first two filters (biocultural filter and disposition filter) are very similar to the levels and filters expected when working with archaeological mortuary samples. This part of the figure borrows heavily from Hoppa's (1996, 1999) graphical depiction of the filter model and is supplemented by key aspects of the NBS. In these first three levels many of the reference collections mimic a highly specialized archaeological mortuary sample such as a crypt sample. For a number of context-specific reasons, the disposition of the deceased will vary, and not all the individuals who die (mortality sample) end up in the same burial or entombment (mortuary sample). Furthermore, not all the individuals who are in the mortuary sample will be recovered for study, or if they are recovered, may be too fragmentary for analysis.

In the living population, not everyone has the same likelihood of dying at a given age and the probability of dying is not based solely on biological issues (Hoppa 1996, 1999; see also Stodder 2008; Wood *et al.* 1992). The NBS emphasizes the importance of the interaction of biological and cultural factors that affect morbidity and mortality (Levins and Lewontin 1998). Poverty, restricted access to resources (when healthy or afflicted with an illness), difficult working conditions, exposure to infectious diseases, and the synergistic interaction of these and other factors may increase morbidity and mortality in different cohorts in some segments of a specific community.

As with archaeological cases, the mortality sample for reference collections is shaped by various socio-economic, gender and religious factors that have had a filtering effect on who is more likely to die at a given age (1. biocultural filter), how the remains are treated and who is included in various types of burial (2. disposition filter) which can take the form of a communal ossuary, elite crypt, dissection hall cadaver, skeletal collection, *Body Worlds* display, etc. These two filters are often cited as having the greatest impact on the composition of various reference collections. However, the effects of the filters are neither simple nor static. For example, different socio-economic and political issues have had various effects on the Terry Collection at different points in the collecting process. Early in the collection period, the poorest individuals, those who lacked social or economic supports in the community at the time of death, were transferred from various institutions in Missouri to Washington University in St Louis for use in anatomical instruction (for more information about the selective sourcing of cadavers in the 19th and early 20th century see Breeden 1975; Harrington and Blakely 1997; Lassek 1958; Muller *et al.* 2017; Wilf 1989). After the mid-1950s, there were two major contrasting socio-economic issues. First, as in the inter-war period, in the post-WW II period, individuals without economic or social supports at the times of their death were still being used for anatomical instruction but in a diminishing number because the economic situation had improved after World War II in Missouri to the point where very few cadavers were available from this source (Trotter 1981). At the same time, changes in social views towards dissection and anatomical instruction along with changes in legislation and regulation allowed for the bequeathing of human remains for anatomical instruction in Missouri (Trotter 1981). In contrast, collecting for the Hamann-Todd Collection ended in the late 1930s with Todd's death, prior to the sweeping legal and social changes related to views of anatomical instruction or major civil rights movements in the United States.[4] Despite the superficial similarities in the sources of skeletons as anatomy hall cadavers, the patterns of variation in various anatomically-derived reference collections should not be expected to be the same. Direct comparisons in the patterns of sexual dimorphism and/or age-related changes between anatomically-derived collections illustrate some of these differences (Albanese 1997a, 1997b, 1997c; Sharman 2013; see Chapter 5, this volume, for examples).

---

[4] In the United States and Canada, these matters fall under state and provincial jurisdiction, respectively. See Chapter 3 which focuses on the Grant collections, but addresses some of these issues for anatomical collections in general in Canada and the Unites States.

When the Terry Collection is compared to a cemetery-derived reference collection, such as the Coimbra Collection, the importance of unraveling socio-economic issues and their impact on variation in reference collections becomes more obvious and particularly important for assessing patterns of variation. As with most of the anatomical collections, the Coimbra Collection clearly consists of individuals of the low socio-economic status in the community (Cunha 1995; Santos 2000). However, it is very misleading to assume that what may be summarized in a single indicator (low socio-economic status) has the same effects on health and skeletal variation, first, through time in St. Louis where the Terry Collection was amassed; or second, in Coimbra as compared to St. Louis. Variation in the effects of differential access to resources should be expected since the network of power relations are 'uniquely configured, socially and historically, in particular places at particular times' (Roseberry 1998:81).

Roseberry (1998) is critical of a comparative approach where historical context and power relations are distilled out so that the basic elements that are considered to be distinctive of the 'type' can be identified and compared through time and space completely devoid of context. Although Roseberry's critique is not directed at skeletal biology or research involving reference collections, many of her criticisms apply to this research. For example, focusing only on the distilled essential elements of the types such as '19th century White male' or '20th century Black female' with no context obscures the underlying processes responsible for what only *appear* to be racial differences and/or secular changes. The observed data, the presence or absence of specific features, is accepted as the reality rather than as tangible results of underlying interactions of biological, historical and economic processes. One of these processes that may explain some of the differences between the Terry and Coimbra Collections relates to the depth of the poverty across multiple generations (see Gowland 2015 for a theoretical framework that addresses some of these issues). The skeletal data support this approach. When looking at secular changes using skeletal data from the Terry Collection and the Coimbra Collection, the pattern of change in long bone length over time varies by collection and not by racial categorization. The 'Whites' of European ancestry and 'Blacks' of African ancestry in the Terry Collection follow one pattern and the Coimbra sample (selected to include only European-born individuals or 'Whites') follow another. This cross-generational impact of stress is so great in the Portuguese collection that *males* from the Coimbra Collection have a mean femur length that is not statistically significantly different than the *females* in the Terry Collection, regardless of racial categorization when birth cohort is controlled for. The 'type' of '19th century White/European male' is meaningless as a category for human variation and useless for assessing secular changes (see Albanese 2010 for the full analysis).

The differences in the impacts of poverty are also compounded by the timing of the collecting. About half of the over 1700 individuals currently in the Terry Collection died during the Great Depression (1929-1939) and may not necessarily have lived in poverty, or at least extreme poverty, during their growth period. The poverty may have been acute and lasted as little as a few months or a few years before death and had a direct impact on the disposition of the body (filter 2 in Figure 1) and less of an effect on the biocultural impacts on morbidity and mortality (filter 1 in Figure 1). There are individuals in the Terry Collection who would have been considered tall at the beginning of the 20th century and would still be considered tall at the beginning of the 21st century. In contrast, the individuals in the Coimbra Collection were from poor families who were unable to pay for the extended care of the remains. These families did not have the resources for expenditures across several generations. The poverty was not only chronic throughout the individual's lifetime but also spanned several generations. Investigating and unraveling the socio-economic issues in the context of the collection period can be very useful in setting parameters for sampling the reference collections in order to maximize representativeness of the sample to address a specific research question. Given the example above by Cunha (1995), it would seem that including a sample from the Coimbra Collection is highly problematic for the development of modern forensic methods (see Ferreira *et al.* 2014). However, including these stressed individuals with highly compromised growth from the Coimbra Collection along with samples from other sources, it is possible to construct a study sample that includes a wide range of human variation, which can be used for various purposes including the development of widely applicable forensic methods.

Some selectivity by anatomical instructors certainly had an effect on the age distribution of the anatomy-derived collections. Children are considered less than ideal for the instruction of gross anatomy because various systems (such as the lymph system) are not fully developed (see also Chapter 3, this volume, where individuals with over-represented ages were removed from the Grant Collection). However, as is the case in archaeological mortuary analysis (for example, see Cannon 1995), gender issues are connected to economic and social factors and have had an impact on the sex ratio in the collections. However, with reference collections derived from anatomical sources, social views of dissection and anatomical instruction have also influenced the final sex ratio in these collections. In most jurisdictions in Canada and the USA anatomical dissection was legal and grave robbing had all but stopped by the end of the 19th century (Terry 1940).[5] However the social stigma associated with being dissected persisted into the first decades of the 20th century in Britain (Richardson 2001) and the United States (Harrington and Blakely 1997). Over the course of the 19th century, anatomical dissection went from being a punishment after death reserved for the most heinous crimes to being a punishment for poverty (Richardson 2001). As in the grave robbing period, individuals of lower socio-economic status continued to be the source of cadavers for anatomical instruction in the first half of the 20th century. Additionally, as in the grave robbing period, community social support networks and political power also had an impact on who was dissected (Breeden 1975; Harrington and Blakely 1997; Lassek 1958; Muller et al. 2017; Wilf 1989). These views of dissection likely had an effect on the sex ratio in the anatomical collections after dissection was legalized. Dissection was considered an indignity that was too horrible to be inflicted even on poor White women (Ginter 2001; Wilf 1989). It is likely that popular perceptions of dissection rather

---

[5] Most of the anatomy acts in Canada and the U.S.A. which were passed in the middle and end of the 19th century are virtually identical or directly influenced by the anatomy act first passed in England in 1832. There are many similarities across these jurisdictions during the grave robbing period (which ended at different times in different places with the passing of an anatomy act) and the period when the procurement of cadavers was legal. See Blake 1955, Richardson 2001; Wilf 1989; Harrington and Blakely 1997; and Chapter 3, this volume.

than medical selectivity had an influence on the sex ratio of cadavers available for dissection. Trotter (1981) notes how difficult it was to make special arrangements to have at least one or two female cadavers in each anatomy class. After World War II, there was a major shift in popular view of human dissection and the regulations were in place for the bequeathment of human remains (Hunt and Albanese 2005; Pregaldin 1958). With these social and legal changes came a shift in the demography and economic status of the available pool of cadavers for anatomical study (Hunt and Albanese; Overholser *et al.* 1956), since dissection was viewed as something that could be done regularly to others besides poor, transient men who were Black or recent immigrants.

From Available Sample to Observed Sample

There are two major types of identified skeletal collections derived from cemetery sources. In the current context it is useful to describe one type as '*post hoc*', or 'after the fact' reference collections and the other as 'constructed' specifically as reference collection. The *post hoc* group includes collections such as St Thomas' (Saunders *et al.* 1995), St Bride's (Scheuer and Bowman 1995) and Spitalfields (Molleson 1995) because they were originally considered archaeological collections with complementary documentary data. Their significance as identified reference collections emerged as they were used for purposes other than reconstructing the original population, such as testing age and sex estimation methods. Various socio-economic and political factors at the time of internment and preservation and taphonomic factors had a direct impact on the make-up of the collection, but these collections were not specifically created as reference collections. Thus, the feedback loops illustrated in Figure 1 do not apply.

The constructed cemetery-derived reference collections are very similar to the anatomical collections and the proposed model has proven useful for research involving these collections (For example, Albanese 2003a, 2010, 2013; Albanese *et al.* 2005,). As with the anatomical collections, the constructed cemetery collections were specifically and systematically created to facilitate research that could only be conducted using a combination of documentary and skeletal data. Because of various social, economic and political reasons, many cemetery-derived reference collections were amassed or are being amassed in Portugal (See Cardoso 2006; Ferreira *et al.* 2014 and Chapter 7). One of the most studied of these Portuguese collections, which is contemporaneous with the Terry Collection, is the Coimbra Collection (Cunha 1995; Rocha 1995; Santos 2000).

With cemetery-derived reference collections such as the Coimbra Collection, the available sample includes all the individuals who were to be interred in the common burial ground (or equivalent) of the cemetery. With reference collections derived from anatomical sources (Terry, Hamann-Todd, Grant Collections), the available or mortuary sample is the pool of cadavers in the morgue used for anatomical instruction. From this available sample, whether it is the common burial ground or the morgue, individuals were selected for inclusion in the reference collection based on criteria such as age at death, sex and geographic origin, which varied over the period of collecting. Below this level, a linear hierarchical model derived from cemetery studies is not the best way to approach the investigation and identification of bias because there are feedback loops caused by the researcher/collector, discipline, and society of the researcher (rather than the society of the study subject). The construction of the collection reinforces perceptions of human variation and the same time is shaped by those concepts. This major deviation from the cemeteries studies model beginnings with filter 3, the collecting filter, and continues with filter 4, the theory and method filter. These filters had a huge impact on which skeletons were included in the collection and what documentary data were verified and readily curated with the collection, and subsequently how the collection was sampled and studied.

A good example of the impact of the collecting filter and the feedback loop is the constructed demographic composition of the Coimbra Collection and the verification and curation of documentary data. Collecting of skeletons occurred in two major phases (Rocha 1995; Santos 2000). In the first phase, females outnumbered males and there were a disproportionate number of females over 60 years of age. In the next phase, first, more males were collected to balance the overall sex ratio; and second, for both males and females, but more so for females, there was an effort to include females under about 60 years of age. In the final composition of the collection, there are equal numbers of males and females in almost each decade of life; there are 47 males and 49 females 60 years or older; and the age range for both sexes is almost identical: 7 to 95 years for females and 7 to 96 years for males.

For the Coimbra Collection, detailed nativity data (place of birth) are listed at the top of the identity form associated with each skeleton. These nativity data are presented at three levels: district, municipality and parish. The prominence of the nativity data may be related to the original motivation for collecting skeletal material at the University of Coimbra (see Chapter 5, this volume). However, as Santos (2000) clearly demonstrates, modern research questions do not have to be limited by the filtering of documentary information that occurred when the collections were amassed (see also Santos and Roberts 2001; Santos and Roberts 2006). Simply because nativity was of great importance 80 years ago does not mean it must limit the research for 21st century anthropologists.

One of the best examples that encompasses both the collecting filter and the theory and methods filter, and illustrates the feedback loops is Mildred Trotter's (1981) efforts to balance the sex ratio in the Terry Collection. Because of the popular views of dissection at the time when many anatomical-derived reference collections were amassed, the males greatly outnumber females by a huge proportion (Hunt and Albanese 2005). The ratio of males to females in the Hamann-Todd collection may be as high as 15:1 (Krogman and İşcan 1986) because the skeletons of almost all the cadavers, predominantly males, used in anatomical instruction were retained by Todd and were included in the skeletal collection. Similarly, in the Grant Collection males initially outnumber females in a ratio of 10.5:1 before the collection was significantly modified to a ratio of about 6.8:1 (see Chapter 3, this volume). In contrast and largely through Trotter's efforts after Terry's retirement in 1941, the Terry Collection was modified in structure and has a very different sex ratio of 1.4:1, with males outnumbering females to a much lesser degree. The obvious positive and less obvious negative significance of Trotter's efforts to include females in the collection cannot be overstated.

It is not a coincidence that Trotter's name is virtually synonymous with stature estimation and that she and Gleser developed race-specific stature estimation equations and used a racial approach to investigate secular changes (Trotter and Gleser 1951, 1952). Trotter's personal interests were clearly affected by her time with Terry, the politics within an anatomy department that had been moving away from research in physical anthropology for decades and a society moving towards a wider acceptance of the use of human remains for research and education (Hunt and Albanese 2005; Chapter 3, this volume). Furthermore, within the discipline of physical anthropology, some researchers (for example, Giles and Elliot 1963) had lamented the lack of availability of 'White' females in the Terry and Hamann-Todd Collections. As it became possible to include bequeathed individuals in the collection, departmental resources were not available to expand the Terry Collection at the same rate as the pre-World War II period, but Trotter did correct what she and others saw as a major imbalance in the collection. Collecting under Trotter's direction shifted from retaining almost every skeleton in the 1930s to a focus on predominantly White females, from the 1940s until the mid-1960s when the collecting stopped with her retirement. Soon after collecting ceased, the collection was transferred to the Smithsonian Institution's National Museum of Natural History where it has been readily available for research ever since.

As with the Coimbra Collection, certain documentary data were prominently curated with the skeletons from the start of collecting. The 'Morgue Record' is the crucial document in the Terry Collection because it accompanied the skeleton throughout the processing and cataloguing of each cadaver/skeleton, and is still curated with the collection (see Hunt and Albanese 2005 for detailed description of this and other key documents). The first line of the morgue record is devoted to information that had to be perpetually kept under the anatomy act in Missouri.[6] The first thing that is listed after meeting the legal requirements was race, which was assigned at the time of collecting (see Chapter 3, this volume, for a contrasting approach used in a different context in Toronto, Canada and New York, USA). Thus, with the majority of the Terry Collection, race was assigned based on early 20th century concepts of human variation that predated the adoption by physical anthropologists of Darwinian evolutionary theory and Mendelian genetics (see Washburn 1951,1963 for critiques of the concepts and approaches that persisted in the discipline well into the 1960s). The race issue is even more complicated because the collecting filter was not static. Collecting for the Terry Collection was directed by two different collectors with different access to resources over several decades during a great deal of social change in the USA. Within the discipline, views regarding the significance of race changed over the collection period: from Hrdlička (1925) to Schultz (1930); to St. Hoyme (1957), and Thieme and Schull (1957). The ranking of races was dropped by the post-World War II period, but the racial assumptions about variation were not. The filters were different for each sex and 'race,' and they varied through time. The criteria for defining 'White' female and

---

[6] Qualified institutions that are legally allowed to accept cadavers for anatomical instruction must retain in perpetuity basic identification information of the deceased. Details vary by jurisdiction in the USA and Canada, but typically include the deceased's name, age, unique identification number (cadaver/skeleton number) and place of death. See Chapter 3, this volume for more information about the anatomy acts in Canada and the U.S.A.

reasons for including those skeletons in the collection in 1921 were very different than in 1951.[7] Trotter's efforts did correct for the sex bias in the collection as is evident in the sex ratio. However, she also introduced new biases to the collection since females and males, and racially defined groups are not matched for year of birth (secular change effects) and year of death (mortality bias). The skeletal variation associated with 'White female' did not remain consistent, and what has been misinterpreted as 'racial' differences are actually the result of mortality bias masked by the collection process (Albanese and Saunders 2006; see also Chapter 5, this volume).

The fourth level, referred to as the skeletal collection sample in Figure 1, influenced and reinforced the collecting filter that significantly affected its structure in the first place. The under representation of 'White' females and not just females was noted by researchers who used prominent anatomically-derived skeletal collections (Giles and Elliot 1963) while Trotter was still actively adding skeletons to the Terry Collection. Furthermore, this same racial structure of the collections had an influence on the theory and methods filter. The basic methodologies for selecting a sample for research involving, for example, sex and age estimation methods with divisions of samples into racial groups have changed little since the Terry and Hamann-Todd Collections became available for research. 'Race' is prominently documented for each individual along with other demographic variables, and thus, it has been assumed for decades that it must be an important source of variation, and that it must be considered and/or controlled. In the late 20th century, racial approaches to research still accounted for a large proportion of the publications in major journals such as the *American Journal of Physical Anthropology* (Cartmill 1998), and many physical anthropologist still consider race to be a meaningful biological concept (Wagner *et al.* 2016). The current methodologies reinforce the research bias and the cultural bias with a feedback loop from the observed sample back to the theory and methods filter. In other words, the structure of the collection mirrors the social structure of the society of the collectors, and the structure of the collection has helped to maintain that social structure through the pursuits of physical anthropologists who have used the collection. A long history within the discipline of selecting samples by 'race' reinforces the idea that it is relevant and must be considered in future research. The collections and the demographic data that were retained have a feedback loop and continue to set some of the parameters of research. In a cyclical fashion, the reference collections, in conjunction with anthropological training and views about race, have limited how human variation is investigated and have greatly reinforced the underlying assumptions that researchers bring with them when using the collections. The relatively recent changes in the discipline to use ancestry instead of race have not corrected any of the problems (see below).

---

[7] The absurdity of these classification systems can best be illustrated with Hrdlička's (1925) detailed criteria for being an 'old American'. The only 'true Whites' that could be considered 'real' Americans had to have been third generation American-born with all four grandparents having been born in the U.S.A. and of western European 'stock'. He excluded himself with these criteria from being part of the 'elite' since Hrdlička was a first generation immigrant from what is now the Czech Republic. Detailed descriptions of changes in scientific stereotypes in the first six decades of the 20th century are beyond the scope of this chapter and have been covered in detail elsewhere (for excellent detailed summaries and critiques see Blakey 1987; Armelagos *et al.* 1982).

Discussion: Examples of the Efficacy of this Approach

Methods for Estimating Sex

Almost all metric sex estimation methods published after World War II are directly or indirectly derived from research by Washburn[8] (1948) and Thieme and Schull (1957). These two papers set a typological methodology for developing sex estimation methods: (1) with few exceptions only race and sex data are considered at the exclusion of all other documentary information available for reference collections; (2) discriminant function analysis is used when multivariate statistical approaches are required; and (3) race is the critical criterion for defining the sample (or dividing the sample into sub-samples), and the accuracy of the method is presented by race (Albanese 2003b). By the beginning of the 21st century metric methods were consistently presented as group-specific and morphoscopic methods were presented in a way that implied they were more widely applicable (Albanese 2003a, 2003b).

The theoretical model described in this chapter was first applied to sample the Terry and Coimbra Collections to develop a general methodology for developing new sex estimation methods that are not population-specific and to develop a specific metric sex estimation method using the os coxa (hipbone) and femur that has high allocation accuracy when applied in modern forensic cases (Albanese 2003a). With the proposed model, the Terry and Coimbra Collections were sampled to construct a study sample that included a wide range of human variation (see also Chapter 5, this volume). Furthermore, various widely held assumptions (filter 4 in Figure 1) about the use of discriminant function analysis and racially defined groups were challenged and alternative approaches were used. The result was a series of multivariate logistic regression equations for estimating sex. This probabilistic method performed very well when tested on a large independent sample including a modern forensic sample from the FDB. Accuracy always exceeded 90% even in cases where the pubic bone data were missing and was as high as 98.5% with complete data, with little to no difference in allocation by sex. Furthermore the method works on problematic cases when morphoscopic methods such as the ventral arc should have worked but provided erroneous results.[9] More recently, tests using a wide range of samples from multiple sources demonstrate that the method consistently provides good results and outperforms other methods (Cardoso *et al.* 2015; Jones and Albanese 2013; Sharman 2013). Follow-up papers expanded the approach to include additional equations to

---

[8] Washburn (1951, 1963) strongly critiqued the reductionist, descriptive approaches to research, including racial concepts that were the norm in physical anthropology into the 1960s. He forced the discipline to conduct research within the framework of evolutionary theory, i.e. the 'New Physical Anthropology' (Tuttle 2000). However, his 1948 paper on sex differences in the pubic bone was essentially the first modern method for sex estimation, and although it was not intentional, this paper in a subtle, yet significant, way set the racial/typological approach for almost all metric sex determination methods that followed (Albanese 2003b).
[9] The conventional wisdom in the discipline has been that morphoscopic methods (assessment of shape) such as the presence or absence of the ventral arc on the pubic bone are applicable across populations or groups, whereas metric methods are group-specific. Independent test of the ventral arc have shown that it is not necessarily universally applicable (see Albanese 2003a for a synthesis and critique of the 'metric versus morphoscopic' debate).

deal with fragmentary os coxae (hip bones) and to develop a series of equations for estimating sex using the upper limb which were tested with independent samples from the Lisbon Collection, Grant Collection and the FDB (Albanese 2013; Albanese *et al.* 2008).

Assessing Patterns of Human Variation and the Study of Past Populations

In the next application of the model (Albanese *et al.* 2005), samples from the Lisbon Collection and Coimbra Collections were used to assess size-related patterns of sexual dimorphism and for the development of an effective sample-specific protocol for sex estimation in cases where remains are damaged and/or commingled. The test case involving a large sample from St Thomas' Anglican Church cemetery in Belleville, Ontario, Canada, which was used in the mid-19th century by European immigrant to Canada (see Saunders *et al.* 1995 for more information).

The same sample that was used for sex estimation methods was also used to critically evaluate the methodology for the study of secular changes using reference collections and to assess secular changes in the Terry and Coimbra Collections (Albanese 2010). The research showed that racial approaches used since Trotter and Gleser (1951) failed to capture or control for patterns of secular change, and in fact the racial approach masked actual patterns of variation across generations. Second, the research demonstrated that 'low socio-economic' status can have very different impacts on human variation in various places and through time. As mentioned above, the Coimbra Collection males' growth was so compromised by the cross-generational impacts of extreme poverty that their mean femur length is not statistically significantly different than the females (regardless of 'race') in the Terry Collection.[10] Third, although age at death and year of birth data can be useful for sample selection to address various research questions, there is no inherent meaning in these data. The age at death data and year of birth day are useful for the study of secular changes only when socio-economic, political and historical contexts are considered.

Methods for Estimating Stature

Most recently, this approach has been used to assess the state of stature estimation methods in forensic anthropology and to develop new equations using various long bones (Albanese *et al.* 2012, 2016a, 2016b; Cardoso *et al.* 2016). Stature estimation is a topic that is considered relatively simple and straight forward and has been considered theoretically and methodologically 'solved' in forensic contexts. The general approach has been to follow Trotter and Gleser's (1952) methodology. Group-specific methods are developed with almost no consideration for the source of the skeletal data, the quality of the stature data or the parameters for defining a group. The sample from the Terry Collection first used to develop sex estimation methods was modified and

---

[10] Femur length as a proxy of overall stature is only one non-specific indicator of health. The physiological impacts of poverty, social inequality and discrimination on growth, development, overall health and aging are complex and varied over the course of a lifetime. For more on these issues, see Alves Cardoso et al. 2016 who studied large samples the Lisbon and Coimbra Collections, and de la Cova 2010, 2011, 2012 who used samples from several of the major American anatomical collections including the Terry Collection.

expanded, and along with samples from the Lisbon Collection and the FDB, we were able to substantiate some conclusions that are in contrast to widely held concepts regarding stature estimation.

First, there are few large skeletal collections that have accurate and verifiable stature data (Cardoso *et al.* 2016). Thus, while there has been much discussion about the relationships between cadaver length, living stature and forensic stature, there is remarkably little evidence supporting the correction factors that have been prescribed for decades. By considering the Terry Collection within the theoretical model, we found that uncorrected stature data from the Terry Collection provided the best estimates of stature when tested using an independent diverse sample (Albanese *et al.* 2016a).

Second, group-specific methods are much more difficult to apply because an unknown must be assigned to a given group before the 'right' equation can be applied, but these methods do not provide any increase in accuracy, precision or reliability (Albanese *et al.* 2012, 2016a, 2016b). Because of the arbitrary nature of these categories, there is no benefit to using equations that are specific to a racial group, nationality, century of birth or any combination of these criteria for group membership.

Third, new generic equations that do not require an unknown to be allocated to a specific group provide the best results most often when estimating stature (Albanese *et al.* 2016b). Independent tests using data from the FDB consistently support the utility of these equations for providing stature data that would be useful in a forensic investigation. Furthermore, in a head to head test with group-specific equations (Mendonça 2000), the new equations based on the sample from the Terry Collection work as well as 'Portuguese-specific' equations when tested on a Portuguese sample from the Lisbon Collection (Albanese *et al.* 2016b).

Fourth, intuitively, theoretically and practically it seems obvious that the femur, as the single largest bone that contributes to standing height, provides the best estimate of stature (Krogman and İşcan 1986), and not surprisingly our results are consistent with this expectation. However, our results across multiple papers and analyses showed that the humerus can be as good as or better than the tibia when estimating stature (Albanese *et al.* 2012, 2016a, 2016b). Small differences in the explanatory power of the bones (based on adjusted r^2) and minor variations in the standard estimate of the error did not translate into better results for the tibia over the humerus. Furthermore, adding the humerus to any multivariate equation increased the accuracy and precision of the estimate of stature. An equation that includes the femur and the humerus will outperform an equation that uses only the femur; an equation that includes the femur, tibia and humerus will outperform an equation using only the femur and tibia; and so forth.

Beyond Race and Ancestry

A critical assessment of typological or racial approaches to research is a theme that has spanned much of the research involving this theoretical model for using reference collections[11]. The model has been used to critique the race concept in three general areas: 1) the use of racial categories for the study of human skeletal variation; 2) the development of race-specific methods for forensic applications; and 3) race or ancestry estimation methods.

The first two areas have been described above. Racial approaches for the study of secular changes provide misleading results. Race-specific forensic methods for estimating stature and sex are more difficult to apply and do not improve accuracy, precision and certainty in a case involving an unknown individual. There are many comprehensive and substantive critiques of the racial and typological approaches used by physical anthropologists beginning over 100 years ago with research by Boas (for example, Boas 1912). However, there is a persistence of a racial or typological approach for investigating human skeletal variation or for the development of group-specific forensic methods. What the critiques of race have lacked was an alternative to the racial approach, particularly in forensic contexts. Now there are alternatives that are demonstrably better than a racial approach for estimating sex and stature regardless of how a forensic anthropologist consider race. If race is part of personal identification in some jurisdictions (for example, Kennedy 1995; Sauer 1992), then that very problematic question (Albanese and Saunders 2006, Komar and Buikstra 2008) can be addressed separately with no impact on other aspects of a biological profile.

The third area is the direct critique of race estimation methods and the problematic fundamental assumptions about race and ancestry in forensic anthropology (Albanese and Saunders 2006). The model presented in this chapter was applied to this issue of race estimation to address issues related to both skeletal variation and historical context for the documentary data that accompanies each skeleton. We found that most race/ancestry estimation methods are based on very small samples, and even when they are based on a larger sample, they perform poorly when tested using large independent samples. Furthermore, regardless of what data are used - metric, morphological, non-metric, anthroposcopic or genetic - we found that forensic anthropologists have used and continue to use concepts of race or ancestry that are not consistent with human variation (see also, Armelagos and Goodman 1998).

One of the problems has been a misinterpretation of human variation when using reference collections. What has been described as racial variation is actually due to another variable such as mortality bias and/or secular changes (See Albanese and Saunders 2006 and Chapter 5, this volume). Another problem has been the misperception of human variation. For example, the cranial index has been foundational in describing racial variation for over 150 years (see Blakey 1987 for additional historical context). The cranial index is calculated as cranial breadth divided by cranial length multiplied by 100, and traditionally index scores have been grouped in categories of cranial shape that range from long crania to hyper-round crania: dolicocranic, up to 75; mesocranic, 75–79.9; brachycranic, 80–84.9; and hyperbrachycranic, 85 or greater (Olivier 1969). Blacks have been described as dolicocranic (France 2001; Gill 1998; Krogman and İşcan 1986; Olivier

---

[11] See also Chapter 9, this volume, for additional context.

1969; İşcan 1981; Rogers 1987; Skinner and Lazenby 1983). Whites have been variously described as dolicocranic (Skinner and Lazenby 1983), mesocranic (France 2001; Gill 1998), both dolicocranic and mesocranic (Rogers 1987), brachycranic (İşcan 1981), and as spanning the mesocranic and brachycranic categories (Krogman and İşcan 1986; Olivier 1969). Using a sample of over 500 individuals from the Terry and Coimbra Collections, we found that the mean for Black males (74.6) is only marginally dolicocranic, and the mean for Black females (76.0) is mesocranic. The means for the European-born Coimbra Collection females (74.4) and males (73.4) are in the dolicocranic range, whereas the means for the Terry Collection White females (77.5) and males (77.5) are in the mesocranic range. Whites from the Terry Collection have more in common with Blacks from the Terry Collection than they do with European-born Whites from the Coimbra Collection. When the cranial length and cranial breadth are presented in a scatter plot, there is complete overlap for Blacks and Whites when cranial variation is compared using racial categories. Any attempt to use what *appear* to be racial characteristics for race estimation in a forensic context can only lead to erroneous results.

Lastly, as Armelagos and Goodman note,

> 'Race, we conclude, has failed to work as a core anthropological concept; it fails to describe and explain variation. What race does do and does well, is type individuals, and this typing supports the existing structures of power.' (1998:259)

In response to this and other critiques about the race concept in physical and forensic anthropology, beginning in the late 1990s and early 2000s, there was a trend towards dropping race-related terminology and replacing those terms with ancestry and continental origin terms.

Unfortunately, a shift in concepts did not accompany the shift in terminology, and what is essentially a typological approach persists in the study of human skeletal variation. If the historical context for the Terry Collection is not considered under the proposed model, the change in terms only obfuscated the underlying highly problematic racial approach. The people who are more recently referred to as 'African' or 'African-American' and were previously referred to as 'Black' from the Terry Collection are actually described as 'Negro' in the original documents. Changing the terms does nothing to change the underlying assumptions made by physical anthropologist in the 1930s when most of the collection was amassed (see Albanese and Saunders 2006 for more details). These terms reflect the *then* current view of human variation. Because someone, even the most recognizable figures of the discipline, wrote something down over 80 years ago does not mean we should accept it without question. They, like all researchers, are not working outside of history and were greatly affected by the context in which they lived.

Regardless of which terminology is used, the continued application of 21st century technology to 19th century concepts of human variation is failing physical anthropology. As Armelagos and Van Gerven (2003:62) note, we have a responsibility as anthropologists to challenge the *status quo* using 'an interdisciplinary and intradisciplinary approach... [and]...reclaim skeletal biology as the means to understand morphology from a functional perspective and adaptation and evolution from a biocultural perspective.'

Conclusion

All skeletal collections, including older and newer reference collections, are biased in different ways and to varying degrees. Stripping individual skeletons and/or a collection as a whole of their context and reducing skeletons to ideal types existing outside of time and space, will necessarily lead to erroneous results, even if they are consistent with expectations in the discipline. Whereas beginning with the assumption that all collections are biased and developing strategies to address or exploit those biases can be an effective approach to research. The model for assessing bias in reference collections presented in this chapter is only one way to pursue research using reference collections. Systematic approaches including variants of this model or entirely different approaches may be required for different research questions and/or different collections.

There are two fundamental issues that need to be considered when working with identified reference collections. First, in addition to other filters that have an effect on the nature and size of mortuary samples that are available for analysis, individual researcher/curator bias, discipline bias and bias in the greater society have had an impact on reference collections. Anthropology is done by people that are operating within the context of a discipline and a greater society at a specific point in time, which has had an impact on the composition of a collection and on what documentary data were collected, verified and curated with the collection. The source and nature of the biases vary over time, and are highly influenced by a number of factors including, the research interests of the collectors/curators (e.g. Terry's interest in normal variation and Trotter's interest in stature), the discipline paradigm (e.g. human variation considered in racial categories), and popular views held by the greater society in which the collecting occurs (e.g. views regarding the human body and the acceptability of dissection).The collections shape and are shaped by these biases. Second, collections are biased samples and not populations. A random sample of a biased sample will result in a biased subsample. But biased collections do not necessarily produce biased results. Reference collections do not have to be representative of the population they were drawn from in order to be very useful for research. It is the size and nature of the *observed* sample and the quality of the supporting documentary data, and not necessarily the source of the reference collection that will place limits on research. The reference collections tend to be large and well documented so it is possible to identify the biases, sample around the biases, and construct an observed sample to address specific research questions, provided that a combination of skeletal and documentary evidence are used and considered within their historical context.

References

Albanese, J. 1997a. A comparison of the Terry Collection and the Grant Collection using the head of the femur and head of the humerus: Implications for determining sex. *Canadian Society of Forensic Science Journal* 30: 167.

Albanese, J. 1997b. Similarities and differences between the Terry Collection and the Grant Collection: The implications of collection and sample selection when developing sex determination methods. Paper presented at the *25th Annual Meeting of the Canadian Association for Physical Anthropology*, London, Ontario.

Albanese, J. 1997c. Skeletal variability in recent North Americans: Implication for the development of new sex determination methods. Paper presented at the *25th Annual Meeting of the Canadian Association for Physical Anthropology*, London, Ontario.

Albanese, J. 2003a. A metric method for sex determination using the hipbone and femur. *Journal of Forensic Sciences* 48: 263–273.

Albanese, J. 2003b. *Identified Skeletal Reference Collections and the Study of Human Variation.* Unpublished PhD thesis, McMaster University.

Albanese, J. 2010. A critical review of the methodology for the study of secular change using skeletal data. In C. Ellis, N. Ferris, P. Timmins and C. White C (eds), *Papers in Honour of Michael Spence*: 139–155. London, Ontario Archaeological Society Occasional Publication No 9.

Albanese, J. 2013. A method for determining sex using the clavicle, humerus, radius and ulna. *Journal of Forensic Science* 58: 1413–1419.

Albanese, J. Cardoso, H. F. V. and Saunders, S. R. 2005. Universal methodology for developing univariate sample-specific sex determination methods: an example using the epicondylar breadth of the humerus. *Journal of Archaeological Science* 32:143–152.

Albanese, J., Eklics, G. and Tuck, A. 2008. A metric method for sex determination using the proximal femur and fragmentary hipbone. *Journal of Forensic Sciences* 53: 1283–1288.

Albanese, J., Osley, S. E. and Tuck, A. 2016b. Do group-specific equations provide the best estimates of stature? *Forensic Science International* 261: 154–158.

Albanese, J., Osley, S. E. and Tuck, A. 2012. Do century-specific equations provide better estimates of stature? A test of the 19th 20th century boundary for the stature estimation feature in Fordisc 3.0. *Forensic Science International* 219: 286–288.

Albanese, J. and Saunders, S. R. 2006. Is it possible to escape racial typology in forensic identification? In A. Schmitt, E. Cunha and J. Pinheiro (eds), *Forensic Anthropology and Medicine: Complementary Sciences from Recovery to Cause of Death*: 281–315. Totowa, NJ, Humana Press.

Albanese, J., Tuck. A., Gomes, J. and Cardoso, H. F. V. 2016a. An alternative approach for estimating stature from long bones that is not population- or group-specific. *Forensic Science International* 259: 59–68.

Alves Cardoso, F., Assis, S. and Henderson, C. 2016. Exploring poverty: skeletal biology and documentary evidence in 19th–20th century Portugal. *Annals of human biology* 43: 102–106.

Armelagos, G. J. and Goodman, A. H. 1998. Race, racism, and anthropology. In A. H. Goodman and T. L. Leatherman (eds), *Building a New Biocultural Synthesis: Political-Economic Perspectives on Human Biology*: 359–378. Ann Arbor, University of Michigan Press.

Armelagos, G. J., Carlson, D. S. and Van Gerven, D. P. 1982. The theoretical foundations and development of skeletal biology. In F. Spencer (ed.), *A History of American Physical Anthropology 1930-1980*: 305–328. New York, Academic Press.

Armelagos, G. J. and van Gerven, D. P. 2003. A century of skeletal biology and paleopathology: contrasts, contradictions, and conflicts. *American Anthropologist* 105: 53–64.

Blake, J. B. 1955. The development of American Anatomy Acts. *Journal of Medical Education* 30: 431–439.

Blakey, M. L. 1987. Skull doctors: intrinsic social and political bias in the history of American physical anthropology with special reference to the work of Aleš Hrdlička. *Critical Anthropology* 7: 7–35.

Boas, F. 1912. Changes in bodily form of descendants of immigrants. *American Anthropologist* 14: 530–562.

Breeden, J. O. 1975. Body snatchers and anatomy professors: medical education in 19th century Virginia. *Virginia Magazine of History and Biography* 83: 321–345.

Cannon, A. 1995. Material culture and burial representativeness. In S. R. Saunders and A. Herring (eds), *Grave Reflections: Portraying the Past Through Cemetery Studies*: 3–17. Toronto, Canadian Scholars Press.

Cardoso, H. F. V. 2006. Brief communication: the collection of identified human skeletons housed at Bocage Museum (National Museum of Natural History), Lisbon, Portugal. *American Journal of Physical Anthropology* 129: 172–176.

Cardoso, H. F. V., Marinho, L. and Albanese J. 2016. Relationship between cadaver, living and forensic stature: a review of current knowledge and a test using a sample of adult Portuguese males. *Forensic Science International* 258: 55–63.

Cardoso, H. F. V., Marinho, L., Vandergugten, J., Simon, E. L., Spake, L., McCuaig, M. and Albanese, J. 2015. Can we improve the reliability of sex estimations in archaeological populations by using non-population specific metric methods? Paper presented at the *43rd Annual Meeting of the Canadian Association for Physical Anthropology*, Winnipeg, October 28–31.

Cartmill, M. 1998. The status of the race concept in physical anthropology. *American Anthropologist* 100: 651–660.

Cunha, E. 1995. Testing identification records: evidence from the Coimbra Identified Skeletal Collections (nineteenth and twentieth centuries). In S. R. Saunders and A. Herring (eds), *Grave Reflections: Portraying the Past Through Cemetery Studies*: 179–198. Toronto, Canadian Scholars Press.

de La Cova, C. 2010. Cultural patterns of trauma among 19th-century-born males in cadaver collections. *American Anthropologist* 112: 589–606.

de la Cova, C. 2011. Race, health, and disease in 19th-century-born males. *American Journal of Physical Anthropology* 144: 526–537.

de la Cova, C. 2012. Patterns of trauma and violence in 19th-century-born African-American and Euro-American females. *International Journal of Paleopathology* 2: 61–68.

Ferreira, M. T., Vicente, R., Navega, D., Gonçalves, D., Curate, F. and Cunha, E. 2014. A new forensic collection housed at the University of Coimbra, Portugal: the 21st century identified skeletal collection. *Forensic Science International* 254: 202.e1–5.

France, D. L. 2001. *Lab Manual and Workbook for Physical Anthropology, 4th Ed.* Belmont, CA, Wadsworth Thomson Learning.

Giles, E. and Elliot, O. 1963. Sex determination by discriminant function analysis of crania. *American Journal of Physical Anthropology* 21: 53–68.

Gill, G. W. 1998. Craniofacial criteria in the skeletal attribution of race. In K. Reichs (ed.), *Forensic Osteology*: 293–317. Springfield, Charles C. Thomas.

Ginter, J. K. 2001. *Dealing with Unknowns in a Non-Population: The Skeletal Analysis of the Odd Fellow Series.* Unpublished Master's Thesis. University of Western Ontario.

Gowland, R. L. 2015. Entangled lives: implications of the developmental origins of health and disease hypothesis for bioarchaeology and the life course. *American Journal of Physical Anthropology* 158: 530–540.

Goodman, H. and Leatherman, T. L. (eds) 1998a. *Building a New Biocultural Synthesis: Political-economic Perspectives on Human Biology.* Ann Arbor, University of Michigan Press.

Goodman, H. and Leatherman, T. L. 1998b. Traversing the chasm between biology and culture: an introduction. In A. H. Goodman and T. L. Leatherman (eds), *Building a New Biocultural Synthesis: Political-Economic Perspectives on Human Biology*: 3–41. Ann Arbor, University of Michigan Press.

Harrington, J. M. and Blakely, R. L. (eds) 1997. *Bones in the Basement: Postmortem Racism in Nineteenth-Century Medical Training.* Washington, Smithsonian Institution Press.

Hoppa, R. D. 1996. *Representativeness and Bias in Cemetery Samples: Implications for Palaeodemographic Reconstructions of Past Populations.* Unpublished PhD thesis, McMaster University.

Hoppa, R. D. 1999. Modeling the effects of selection-bias on palaeodemographic analyses. *Homo* 50: 228–243.

Hunt, D. R. and Albanese, J. 2005. The history and demographic composition of the Robert J. Terry Anatomical Collection. *American Journal of Physical Anthropology* 127: 406–417.

Hrdlička, A. 1925. *The Old Americans.* Baltimore, Williams and Wilkins.

İşcan, M. Y. 1981. Race determination from the pelvis. *OSSA* 8: 95–100.

Jones, G. and Albanese, J. 2013. Head-to-head test of Fordisc 3.0 and Albanese 2003 models for sex determination using the hip bone and femur. Paper presented at the *41st Annual Meeting of the Canadian Association for Physical Anthropology*, Scarborough, October 17-20.

Kennedy, K. A. R. 1995. But professor, why teach race identification if races don't exist? *Journal of Forensic Science* 40: 797–800.

Komar, D. A. and Buikstra, J. E. 2008. *Forensic Anthropology: Contemporary Theory and Methods.* New York, Oxford University Press.

Komar, D. A. and Grivas, C. 2008. Manufactured populations: what do contemporary reference skeletal collections represent? A comparative study using the Maxwell Museum documented collection. *American Journal of Physical Anthropology* 137: 224–33.

Krogman, W. and İşcan, M. Y. 1986. *The Human Skeleton in Forensic Medicine.* Springfield, Charles C. Thomas.

Lassek, A. M. 1958. *Human Dissection: Its Drama and its Struggle.* Springfield, Charles C. Thomas.

Levins, R. and Lewontin, R. 1998. Forward. In A. H. Goodman and T. L. Leatherman (eds), *Building a New Biocultural Synthesis: Political-Economic Perspectives on Human Biology*: xi–xv. Ann Arbor, University of Michigan Press.

Marinho, L. and Cardoso, H. F. V. 2013. BoneMedLeg: two new collections of identified human skeletons being amassed in Porto (Portugal) for forensic purposes. Paper presented at the *17th World Congress of the International Union of Anthropological and Ethnological Sciences*, Manchester, UK.

Meindl, R. S., Russel, K. F. and Lovejoy, C. O. 1990. Reliability of age at death in the Hamann-Todd Collection: validity of subselection procedures used in blind tests of the summary age technique. *American Journal of Physical Anthropology* 83: 349–357.

Mendonça, M. C. 2000. Estimation of height from the length of long bones in a Portuguese adult population. *American Journal of Physical Anthropology* 112: 39–48.

Molleson, T. 1995. Rates of aging in the eighteenth century. In S. R. Saunders and A. Herring (eds), *Grave Reflections: Portraying the Past Through Cemetery Studies*: 199–222. Toronto, Canadian Scholars Press.

Muller, J. L., Pearlstein, K. E. and de la Cova, C. 2017. Dissection and documented skeletal collections: embodiments of legalized inequality. In K. C. Nystrom (ed.), *The Bioarchaeology of Dissection and Autopsy in the United States*: 185–201. Cham, Springer International Publishing.

Olivier, G. 1969. *Practical Anthropology*. Springfield, Charles C. Thomas.

Ousley, S. D and Jantz, R. L. 1998. The forensic data bank: documenting skeletal trends in the United States. In K. Reichs (ed.), *Forensic Osteology*: 441–458. Springfield, Charles C. Thomas.

Overholser, M. D., Alexander, W. F. and Trotter, M. 1956. Can Missouri continue the teaching of human anatomy effectively? *Missouri Medical Journal* 53: 474–476.

Pregaldin, A. J. 1958. Comments: property in corpses. *St Louis University Law Journal* 5: 280–297.

Richardson, R. 2001. *Death, Dissection, and the Destitute*. New York, Routledge and Kegan Paul.

Reeve, J. and Adams, M. 1993. *The Spitalfields Project: The archaeology, Across the Styx (Vol. 1)*. York, Council for British Archaeology.

Rocha, A. 1995. Les collections ostéologiques humaines identifiées du Musée Anthropologigue de l'Université de Coimbra. *Antropologia Portuguesa* 13: 7–38.

Rogers, S. L. 1987. *Personal Identification from Human Remains*. Springfield, Charles C. Thomas.

Roseberry, W. 1998. Political-economy and social fields. In A. H. Goodman and T. L. Leatherman (eds), *Building a New Biocultural Synthesis: Political-Economic Perspectives on Human Biology*: 75–91. Ann Arbor, University of Michigan Press.

St. Hoyme, L. E. 1957. The earliest use of indices for sexing pelves. *American Journal of Physical Anthropology* 15: 537–546.

Santos, A. L. 2000. *A Skeletal Picture of Tuberculosis*. Unpublished PhD thesis, University of Coimbra.

Santos, A. L. and Roberts, C. A. 2001. A picture of tuberculosis in young Portuguese people in the early 20th century: a multidisciplinary study of the skeletal and historical evidence. *American Journal of Physical Anthropology* 115: 38–49.

Santos, A. L. and Roberts, C. A. 2006. Anatomy of a serial killer: differential diagnosis of tuberculosis based on rib lesions of adult individuals from the Coimbra Identified Skeletal Collection, Portugal. *American Journal of Physical Anthropology* 130: 38–49.

Sauer, N. J. 1992. Forensic anthropology and the concept of race: if races don't exist, why are forensic anthropologists so good at identifying them? *Social Science and Medicine* 34: 107–11.

Saunders, S. R. and Herring, A. (eds) 1995a. *Grave Reflections: Portraying the Past Through Cemetery Studies*. Toronto, Canadian Scholars Press.

Saunders, S. R. and Herring, A. 1995b. Preface. In S. R. Saunders and A. Herring (eds), *Grave Reflections: Portraying the Past Through Cemetery Studies*: 1–2. Toronto, Canadian Scholars Press.

Saunders, S. R., Herring, A., Sawchuk, L. A. and Boyce, G. 1995. The nineteenth century cemetery at St. Thomas' Anglican Church, Belleville: skeletal remains, parish records, and censuses. In S. R. Saunders and A. Herring (eds), *Grave Reflections: Portraying the Past Through Cemetery Studies*: 93–117. Toronto, Canadian Scholars Press.

Scheuer, J. L. and Bowman, J. E. 1995. Correlation of documentary and skeletal evidence in the St. Bride's Crypt population. In S. R. Saunders and A. Herring (eds), *Grave Reflections: Portraying the Past Through Cemetery Studies*: 49–70. Toronto, Canadian Scholars Press.

Schultz, A. H. 1930. The skeleton of the trunk and limbs of higher primates. *American Journal of Physical Anthropology* 2: 303–438.

Sharman, J. A. 2013. *Age, Sex and the Life Course: Population Variability in Human Ageing and Implications for Bioarchaeology*. Unpublished PhD thesis, University of Durham.

Skinner, M. and Lazenby, R. A. 1983. *Found! Human Remains*. Burnaby, Archaeology Press.

Stodder, A. L. W. 2008. Taphonomy and the nature of archaeological assemblages. In M. A. Katzenberg and S. R. Saunders (eds), *Biological Anthropology of the Human Skeleton. Second Edition*: 71–114. New Jersey, Wiley and Sons.

Tarlow, S. (ed.) 2015. *The Archaeology of Death in Post-medieval Europe*. Berlin, Walter de Gruyter GmbH and Co KG.

Terry, R. J. 1940. On measuring and photographing the cadaver. *American Journal of Physical Anthropology* 26: 433–447.

Thieme, F. P. and Schull, W. J. 1957. Sex determination from the skeleton. *Human Biology* 29: 242–273.

Trotter, M. 1981. Robert James Terry, 1871–1966. *American Journal of Physical Anthropology* 56: 503–508.

Trotter, M. and Gleser, G. C. 1951. Trends in stature of American whites and negroes born between 1840 and 1924. *American Journal of Physical Anthropology* 9: 427–440.

Trotter, M. and Gleser, G. C. 1952. Estimation of stature from long bones of American whites and negroes. *American Journal of Physical Anthropology* 10: 463–551.

Tuttle, R. H. 2000. Obituaries: Sherwood Larned Washburn (1911-2000). *American Anthropologist* 102: 865–869.

Wagner, J. K., Joon-Ho Yu, J. H., Ifekwunigwe, J. O., Harrell, T. M., Bamshad, M. J. and Royal, C.D. 2016. Anthropologists' views on race, ancestry, and genetics. *American Journal of Physical Anthropology* 162: 318–327.

Washburn, S. L. 1948. Sex differences in the pubic bone. *American Journal of Physical Anthropology* 6: 199–207.

Washburn, S. L. 1951. The new physical anthropology. *Transactions of the New York Academy of Sciences (Series II)* 13: 298–304.

Washburn, S. L. 1963. The study of race. *American Anthropologist* 65: 521–531.

Wilf, S. R. 1989. Anatomy and punishment in late eighteenth-century New York. *Journal of Society and History* 22: 507–30.

Wood, J. W., Milner, G. R., Harpending, H. C. and Weiss, K. M. 1992. The osteological paradox: problems of inferring prehistoric health from skeletal samples. *Current Anthropology* 33: 343–370.

Chapter 5

Bioarchaeology and Identified Skeletal Collections: Problems and Potential Solutions

Jennifer Sharman[1] and John Albanese[2,3]

[1] Independent Researcher

[2] Associate Professor, Department of Sociology, Anthropology and Criminology, University of Windsor, 401 Sunset Avenue, Windsor, Ontario, N9B 3P4, Canada

[3] Research Associate, Centre for Forensic Research, Simon Fraser University, 8888 University Dr., Burnaby, BC, V5A 1S6, Canada

Introduction

Documented collections of human skeletal remains are invaluable resources for bioarchaeology and forensic anthropology. In this context, 'documented' means that age at death and sex of individuals in the collection are known from premortem documentary data or other sources of extrinsic knowledge; sometimes, other information is also available, such as names of individuals, their causes of death, and/or occupation in life (though occupation may have changed over the life course; see Henderson *et al.* 2013). Generally, such collections are fairly recent (from approximately the last hundred years), and also fairly rare: curation of skeletons for research requires amenable local laws and/or regulations for the donations of bodies to the relevant institutions, access to skeletons (through donation or excavation of cemetery or archaeological sites), facilities and resources to process the skeletons and verify documentary information, space and permissions to appropriately curate fairly large numbers of skeletons, and a willing and able curator.

Throughout the 19th century and well into the first half of the 20th century, people interested in studying human bones were trained as anatomists and/or medical doctors. Throughout the 19th century they were more concerned with crania than whole skeletons, with a focus on the study of 'racial' differences in order to categorise people and maintain power relations (Blakey 1987; Armelagos et al 1982). By the first decades of the 20th century, physical or biological anthropology began emerging as a discipline, although most scholars still received their training in anatomy departments (Armelagos and Van Gerven 2003; Caspari 2009; Dias 1989). Accordingly, one of the earliest American collections was of crania only – Samuel Morton's collection was intended to educate anatomy students, with curation beginning in Philadelphia, Pennsylvania, USA, around 1830 and continuing until his death in 1852 (Buikstra and Gordon 1981). Sir William Turner (1823–1916), of Edinburgh University, is credited with being one of the first to recognise the comparative value of a documented skeletal collection; his collection was in place when Robert J. Terry was a visiting scholar in 1898 (Tobias 1991). The idea soon spread to anatomy departments in Africa, North America and elsewhere (see Hunt and Albanese 2005; Tobias 1991).

The availability of premortem demographic information for skeletons allows for different types of questions to be asked. For example, standards and methods for age at death and sex estimation can be developed and tested using known age and sex skeletal series. This is not possible with older archaeological skeletal samples that lack premortem records, because there is no way to test skeletal indicators of age and sex reliably on material for which this information is not known. The development of age and sex standards using documented collections are especially important because age and sex estimation are the foundations upon which subsequent bioarchaeological and forensic anthropological interpretation are built, including research on health and disease, (palaeo)demography, diet, mobility, and patterns of morbidity and mortality. Documented collections also provide the opportunity to compare other skeletal traits, reflecting a wide range of normal variation. The study of biomechanics, palaeopathology and growth and development in the past (when non-adult skeletons are also studied) also make use of known age and sex collections. However, bioarchaeology is not the only discipline to benefit from studying documented skeletal collections. These collections have been used widely for research in palaeoanthropology, neurosurgery, orthopaedics and medical implants, medical and dental training and changes in dissection techniques through time (Mitchell 2012, Hunt and Albanese 2005, Tobias 1991).

Documented collections of human skeletal remains are known to have limitations and pitfalls (often, non-adults and/or females are absent or few in number, for instance), and are not generally representative of the population from which they are drawn (Dayal *et al.* 2009; Hunt and Albanese 2005; Komar and Grivas 2008; Usher 2002). Albanese (2003b) and Saunders *et al.* (1995) emphasise the importance of recognising the biases inherent in the collection under study; identification of bias is the first step in appropriate use of data from documented collections (whether cemetery or anatomical in origin). These biases can then be controlled, quantified or even exploited in order to ask more specific research questions (Albanese 2003b; also see Chapter 4, this volume). Documentary and historical records, and their comparison to skeletal evidence, are not only important tools in recognising bias (Saunders *et al.* 1995): the quality of the documentary information for any individual in an identified skeletal collection and for the collection as a whole represents the value of that collection.

Alongside the inherent biases in documented collections of human skeletal remains, there are ethical considerations as well. It is a tenet of biological anthropologists, bioarchaeologists, and forensic anthropologists that human skeletal remains must be treated with dignity and respect in all stages of research. The curation of skeletal remains comes with separate ethical concerns; where individuals have willed their bodies to science, there are fewer concerns, but with archaeological or more modern skeletons excavated from cemeteries, where individuals likely expected their remains to remain, then researchers must think carefully about framing their research in respectful, appropriate ways. If descendants of individuals are uncomfortable with the remains of their ancestors being studied and curated, then repatriation is necessary (in the US, see NAGPRA 1990); other descendants may approve curation, but should be consulted when research involves their ancestors (see, for example, CAPA-ACAP 2015). Where there are no descendants, there may be bias in socioeconomic status in terms of who ends up in a collection, which we discuss later in this chapter. Where skeletal remains are likely those of oppressed or marginalized individuals, Zuckerman and colleagues (2014) suggest framing research questions in terms of giving individuals back their voice; letting their skeletons tell their story, of how oppression and marginalization affected their health and lived experience (see also Chapter 9 this volume). In this way, the researcher gives back to the individuals (or groups) rather than simply taking what they wanted. Professional organizations offer codes of ethics for work involving human skeletal remains (e.g. AAPA 2003;[1] BABAO 2010,[2] CAPA-ACAP 2015[3]), and include other important considerations such as disseminating information to the public. Documented collections are invaluable assets to research on the lives of past humans, which, indeed, should produce results that inform our present human condition; as such, it is vital to work ethically and respectfully with the skeletons of the once-living humans in these collections.

Collectively, we have conducted research involving some of the major identified skeletal and virtual collections from around the world[4]: Coimbra (Rocha 1995), Dart (Dayal *et al.* 2009), Forensic Anthropology Data Bank (Ousley and Jantz 1998), Grant (see Chapter 3 this volume), Lisbon (Cardoso 2006), Pretoria (L'Abbé *et al.* 2005), St. Thomas' Cemetery database (Saunders *et al.* 1995), Spitalfields (Molleson and Cox 1993) and Terry (Hunt and Albanese 2005). In this chapter, we use our experience in working with documented collections of human skeletal remains to provide background information and suggestions that researchers may consider before setting out on their own research using these collections. We identify some of the critical issues that need to be considered when working with identified collections[5] and present some practical solutions to overcome these potential problems to maximize the research potential of these vital resources. A good starting place is to ask the classic five Ws: Who, What, When, Where, Why? Here, we start with 'What'.

What Were the Reasons for Collecting and *What* was the Source of the Skeletons?

The reasons for the collection of skeletons and their source has had an enormous influence on the demographic composition of and the range of skeletal variation in collections. Skeletons can be obtained in several ways: the donation of bodies to research, often first dissected by medical students or used for forensic research, is one possibility; skeletons may also be curated following excavations of church crypts by archaeologists; in some countries where burial is temporary, cemetery staff may excavate burials for

---

[1]   Code of Ethics of the American Association of Physical Anthropologists, 2003. AAPA – American Association of Physical Anthropologists, USA.

[2]   Code of Ethics, 2010. BABAO. British Association of Biological Anthropology and Osteoarchaeology. BABAO Working-group for ethics and practice. Accessed online 26/02/2017 http://www.babao.org.uk/assets/Uploads/code-of-ethics.pdf

[3]   CAPA-ACAP. The Canadian Association for Physical Anthropology/L'Association Canadienne d'Anthropologie Physique. 2015. Code of Ethics. Accessed online 26/02/2017 http://www.capa-acap.net/sites/default/files/basic-page/capa_code_of_ethics_-_oct_2015.pdf

[4]   In this chapter, collection refers to both real accumulations of skeletons and databases.

[5]   Usher 2002 provides a list of identified collections; another list can be found online: Skeletal Collections Database. Accessed online 25/02/2017 http://highfantastical.com/skeletal-collections/

reclamation by relatives, and if skeletons are not reclaimed, curation may occur at that point (Chapter 6 and 7 describes this in the Portuguese context). Some collections originate from archaeological cemeteries, and will have different characteristics in terms of representativeness than collections derived from cadavers retained originally for teaching anatomy (Albanese 2003b).

In some cases, the goal was to curate specific types of individuals and thus these collections (or sub-collections) were not intended to be representative of the population as a whole. Examples of this type of collection include the pathological collection at the Museum of Pathological Anatomy, in Vienna, Austria; the Pathological Anatomy Museum in Rome, Italy; the Galler Collection of pathological cases, at the National History Museum, in Basel, Switzerland; the Standford-Meyer Osteopathology Collection, at the San Diego Museum of Man, California, USA, and the Huntington Collection, at the National Museum of Natural History, Smithsonian Institution (Albanese 2003b; Hunt and Albanese 2005; Rühli *et al.* 2003; Usher 2002).

Anatomical Collections

At the beginning of the 20th century, when physical anthropology was a branch of anatomy, several influential anatomists in key positions as chairs of their respective anatomy departments began retaining the skeletons of the cadavers that were used for anatomical instruction (see Hunt and Albanese 2005 for more information). After dissection by medical students, skeletons might be curated into a documented skeletal collection.

The reasons behind inclusion or non-inclusion of individuals in a documented skeletal collection are similar to those of cemetery collections. Social and cultural biases dictate who is buried in which location (for example, a crypt as opposed to a churchyard) and are essentially the same for cadaver-based collections, except that the biases there may be those of the researcher or curator of the collection, deciding which individuals should be included (Albanese 2003b). These can be socioeconomic factors, sex and/or gender, ethnic background, religion, age, or occupation.

The Dart Collection

The Raymond A. Dart Collection is curated at the Anatomy Department of the University of the Witswatersrand, in Johannesburg, South Africa, and consists of 2605 individuals of known age and sex (Dayal *et al.* 2009). This collection of human skeletal remains was begun in 1923 by Raymond A. Dart, Head of the Anatomy Department at the university. Dart had been inspired by a six month visit as a Rockefeller Fellow to Washington University in the United States, where Dr. Robert J. Terry was collecting human skeletal remains of known age, sex and 'race' (the famous Terry Collection, now curated at the Smithsonian Institution in Washington, DC, USA) (Tobias 1987). Phillip Tobias, Dart's student and successor, named the collection after his teacher and colleague and continued collecting skeletons (Dayal *et al.* 2009). Maciej Henneberg continued curation in the 1980s and 90s, but focused on ensuring more equality in representation of various groups (for example, by sex or different tribal groups). While curation is ongoing, the numbers of skeletons added to the collection annually is less than seen in previous decades due to an increase in demand for skeletons by the medical school for teaching anatomy to medical students (Dayal *et al.* 2009).

Skeletons in the collection began as bodies either unclaimed or donated to the Medical School of the University of the Witwatersrand for dissection by medical students (Dayal *et al.* 2009). There are also some individuals from archaeological sites, victims of mine accidents and forensic cases for whom less information is available (these are not included in the 2605 individuals) (Dayal *et al.* 2009; Tal and Tau 1983). The South African Human Tissues Act (No. 65) of 1983 and previous legislation have allowed for medical schools to procure teaching and research material (Dayal *et al.* 2009).

The Dart Collection, like many other collections originating from unclaimed or donated cadavers, consists of more males than females; while there are 1840 males, only 756 skeletons are female, for an approximate ratio of 3 to 1 (Dayal *et al.* 2009; Tal and Tau 1983). There are also more black South Africans than white South Africans in the collection; South African Asians and Indians and non-adults generally are also underrepresented.

The Pretoria Collection

The Pretoria Bone Collection is another example of an anatomical collection; it is curated in the Department of Anatomy, at the University of Pretoria's Medical School, in Pretoria, South Africa. The collection was initiated shortly after the Medical School and Department of Anatomy were established in 1942 (L'Abbe *et al.* 2005). Reorganisation was undertaken in 2000 to allow easier access for research. Like the Dart Collection, the Pretoria Collection skeletons are either bequeathed or unclaimed bodies. Skeletons are either used for research or for student-teaching, as complete or partial skeletons (L'Abbe *et al.* 2005). Information on skeletons in the student-teaching part of the collection may not be complete, but age, sex, population group ('ancestry' and 'ethnicity'), and date of death is known for individuals in the research collection, and cause of death is listed for most. Curation is ongoing, with 50 to 100 cadavers accepted every year, adding to the research (mainly) and student-teaching subdivisions of the collection (L'Abbe *et al.* 2005). There are 290 complete skeletons, 704 complete skulls and 541 complete postcrania currently available for research (L'Abbe *et al.* 2005). Dates of birth range from 1906 to 1951 (Steyn and İşcan 1999). As in other collections, males are more numerous than females and, similar again to the Dart Collection, there are more black South Africans than white South Africans. The white South Africans in the collection tend to be older than the black South Africans (see Example 2 later in this chapter for discussion on potential causes for this bias).

The Terry Collection

The Terry Collection consists of over 1700 individuals and was collected at Washington University in St. Louis, Missouri, USA. The collection is now at the Smithsonian Institution's National Museum of Natural History. After several false starts, significant retention of skeletons for the Terry Collection really began in the early 1920s, peaked in the 1930s (the decade before Terry's retirement) and continued in a much more selective way through the efforts of Mildred Trotter until the mid-1960s. The age at death range is from 14 to 102, and the year of birth, which was calculated by subtracting age from year of death, ranges from 1828 to 1943. Largely through Trotter's efforts, the sex ratio is atypical for anatomical collections because males only outnumber females at a ratio of about 1.4:1.

Terry and his assistants established a standardised protocol, which varied in the decades of collecting skeletons (Terry 1940; Trotter 1966, 1981). For example, during the 1930s, when most of the collection was amassed, almost every cadaver was measured and photographed in a standing position using specially constructed pivoting tables; as a result, over 900 skeletons have stature data that are accurate, consistent and verifiable through the standardised photographs (Albanese *et al.* 2016; Cardoso *et al.* 2016; Hunt and Albanese 2005).

What remained consistent throughout the collection process was an attention to verifying demographic data. Death certificates and other records had to be kept under the regulations of the Missouri anatomy act and these official records were not altered. But the data curated with the collection were updated as more information about the deceased was confirmed by third parties, and as more information became available about cause of death over the course of the dissection, which in some ways functioned as an autopsy.

Grant Collection

The J.C.B. Grant Collection is curated at the University of Toronto, in Toronto, Canada, by the Anthropology Department (see Chapter 3 this volume for a detailed description). Dr. Grant began curating the human skeletal remains that make up the Grant Collection in 1928. Medical students dissected the unclaimed corpses, and remains were later macerated for addition to the skeletal collection until the early 1950s.

At one point the Grant Collection included the skeletons of over 350 individuals, but currently consists of just 202 individuals (175 are male and 18 female). The majority of individuals are of European origin, and male, and were unclaimed because they were recent immigrants with no relatives nearby, 'transients' or migrant workers (Bedford *et al.* 1993: 288). No information is available on the individuals' lives prior to immigration or life conditions at death. As such, it is difficult to determine whether these individuals were disadvantaged during growth and development or only at their time of death.

Despite the small sample size and high proportion of males to females, the Grant Collection still provides a useful comparison with other identified collections, for some research questions. Researchers hoping to develop age or sex standards or methods would not want to use the Grant Collection alone, but for a test of methods, comparison sample for other anatomical collections, or for testing methods in realistic forensic contexts, the Grant Collection has much value (for example, Albanese 2013).

Cemetery Collections

Cemetery-derived reference collections can be archaeological (used as reference collections *post hoc*) or modern (planned or constructed as reference collections). Identified archaeological cemetery collections may be excavated and curated after modernisation of churches necessitates excavation of an old or forgotten cemetery. Initially, these collections were used to reconstruct the populations from which they were drawn; later, because of the availability of documentary data, the collections started to be used as reference collections (see Saunders and Herring 1995 for several examples). The collection process for modern cemetery collections shares some similarities with anatomically-derived reference collections, and were planned and constructed for similar purposes; one of the main differences is the source of the remains. Regardless of the type of cemetery collection, both archaeological and modern cemetery collections may be unrepresentative due to excavation (partial excavation of cemeteries, or differential burial practices), recovery and preservation biases, but to varying degrees (for more information on these potential biases, see Bello *et al.* 2006; Buikstra and Konigsberg 1985; Hoppa 1999; Scheidel 2001a, 2001b; Walker *et al.* 1988).

Modern Cemetery Collections: The Coimbra Collection

Some cemetery collections are deliberately constructed from a cemetery source for skeletal research purposes, which is the case for the Lisbon and Coimbra collections, both in Portugal. With no perpetual care as in North America and some parts of Europe, in Portugal, the remains are exhumed from the primary burial and the families of the deceased must pay for the care of the remains. If payments cease, bones are cremated or reburied in a communal grave (Cardoso 2006; Wasterlain *et al.* 2009). Instead of cremation or reburial, if no relatives claim the skeleton after a few years, museums may curate the bones (Cardoso 2006; see also Chapter 6).

There are several series of identified skulls and identified full skeletons at the University of Coimbra (Ferreira *et al.* 2014; Cunha 1995; Rocha 1995; Santos 2000). In this chapter, 'Coimbra Collection' refers to the series of 505 identified skeletons known as the *Colecção de Esqueletos Identificados* (Identified Skeletal Collection), which includes individuals who died between 1904 and 1938 and were exhumed from the *Cemitério Municipal da Conchada* (Conchada Municipal Cemetery) in Coimbra (Rocha 1995; Santos 2000; Santos and Roberts 2006; Wasterlain *et al.* 2009). The ages at death range from seven to 96 years, with years of birth from 1826 to 1922 (Coqueugniot and Weaver 2007; see Santos 2000, for additional information about calculating year of birth); while the vast majority were born in Portugal, six were African-born, one Brazilian-born and two were born in Spain (Coqueugniot and Weaver 2007). The recovery of fingernails in some cases attests to the rigour in the recovery and cataloguing of remains for each individual.

Information regarding each individual has been compiled into a record book, including name, age at death, sex, place and cause of death, marital status, occupation (for most; for a discussion of limitations of occupation records, see Chapter 8), name of the parents, birth place and location in the cemetery of original burial. Cause of death information has been correlated with hospital records, autopsy records and other records where possible; a good correspondence between the collection's information and the other records indicate reliable data (Santos and Roberts 2001).

What is known about the collection processes suggests that careful planning went into the assembly of the collection for research and teaching purposes (Santos 2000) and did not result in a random sampling of either the cemetery or the greater population of the District of Coimbra. The demographic data for the collection strongly support this assessment (see Chapter 4, this volume).

As with the anatomical reference collections, the collections at the University of Coimbra would not have been possible without the efforts of a few highly influential individuals who held key positions, and had access to resources and human remains. The first series of skulls was collected by the Director of the Anthropology Section of the University of Coimbra's Natural History Museum, Bernardino Luís de Machado Guimarães, who also served as President of the Republic in 1915–1917 and 1925–1926 (Santos 2000). The second series of skulls and the identified skeletal collections were amassed under the direction of Eusébio Barbosa Tamagnini de Matos Encarnação, who succeeded Machado as the Director of the Anthropology Section, and also served as Minister of Public Instruction (Santos 2000).

Archaeological Cemetery Collections: The Spitalfields Collection

Archaeological cemetery collections have often been created through historical accidents, such as the Spitalfields Collection, London, UK, and the St. Thomas' Collection, from Belleville, Ontario, Canada. As noted above, these collections are often initiated after modern construction work at a church has discovered a forgotten cemetery (perhaps grave markers deteriorated or were removed at some point, or never existed), or a known cemetery or crypt must be moved for construction to continue, stimulating the need for archaeological excavation of the graves and associated remains.

Whether the collection was deliberately constructed or the result of archaeological excavation can impact the state of preservation of the skeletons. If deliberately constructed, it is possible that more complete skeletons, in better condition, are preferentially added to the collection, while cemetery collections that curate the entire excavated cemetery will likely consist of skeletons that are differentially preserved (different burial years result in varying amounts of time spent buried, variation in coffin material may lead to differing preservation of individuals contained therein, etc.). The Spitalfields Collection provides an example of differing preservation within a single-origin cemetery collection.

The skeletal collection often referred to as 'Spitalfields' consists of excavated skeletal remains from the crypt of Christ Church, in Spitalfields, east London, England. Christ Church's crypt was used for interments between 1729 and 1859 (Adams and Reeve 1987); the former was the year of consecration of the church, the latter when burials in the crypt officially ceased, although the last burial to actually take place was in 1852 (Cox 1996). Excavation took place between 1984 and 1986, after a 1981 decision to clear the vaults for installation of necessities such as a boiler room, kitchen and toilets (Cox 1996). The remains are currently curated at the Natural History Museum in London. Spitalfields is a rare example of an identified archaeological collection, as archaeological skeletal populations are not typically of known age and sex; such collections are particularly valuable because of their age (in this case, the 18th and 19th centuries), as they give researchers the opportunity to add the dimension of time within a specific historical context to their research. What is often reduced to a single indicator referred to as secular change can be considered in a much more sophisticated manner by considering detailed context-specific information about geographic, environmental and cultural factors that affect growth and development, normal variation, and morbidity and mortality.

The total number of individuals excavated from the crypt of Christ Church was 987; of these, 600 are unidentified, but 387 are of known name, sex and age at death, due to inscriptions on coffin plates (Cox 1996). The degree of preservation of the individuals was highly variable. While many skeletons were in good condition, some were very fragile and crumbly; as noted by Molleson and Cox (1993: 10), some individuals were represented by '...a sediment of crystal debris', while others still had intact soft tissue (including internal organs, skin and hair) or were naturally mummified. Along with the human remains in varying states of decomposition were some grave goods, burial clothing and coffin textiles, insect remains, fungal blooms, adipocere, and minerals (mostly brushite) (Molleson and Cox 1993; Cox 1996).

Potential Impacts of Curation

In addition to the taphonomic processes that affect cemetery collections, the curatorial process may have an impact in varying degrees on cemetery and anatomical collections. In some cases, the problems arise from errors with curation, which may have an impact on specific individuals in terms of mixing individuals' skeletal elements or mislabelling an individual (i.e. their age or sex), but a careful review of documentary information and skeletal elements can be used to unravel this type of error (for example, see Chapter 3, this volume involving the Grant Collection). In other cases, problems outside of the control of collection curators can also contribute to mixing of skeletal elements; the Dart Collection was unfortunately subjected to such outside influences. The collection, initially stored in the Medical School basement, was flooded in 1959 after pipes burst in the street outside (Dayal et al. 2009: 326). While staff attempted to rescue the 'free-floating' bones, some mixing of individuals occurred. Skeletons were laid out to dry on the roof, and then placed into boxes; however, each skeletal element was not labelled with reference numbers at the time. As a result, some mixing of individuals occurred at this point (Dayal et al. 2009). Work was undertaken in the mid-1980s to create an electronic record, resolve some problems of intermingling, and to deaccession some skeletons (those with damage, or lack of provenience if archaeological or donated/undocumented material). Work is ongoing to overcome these issues, but they still represent a possible source of bias.

Who is in the Collection?

Basic Demographic Parameters: Age and Sex

Addressing the 'Who' question is especially important for formulating research questions in bioarchaeology and forensic anthropology, because age and sex are the essential parameters on which other interpretations or research questions hinge. Considering the collection in terms of its basic demographic parameters (i.e. sex ratio and age at death range/age frequencies) is the first step in determining the appropriateness of a collection for any particular research question. For example, research questions regarding aspects of growth and development clearly require a sample with non-adults (under 20 years of age at death, if using the standards of Buikstra and Ubelaker 1994); thus, when considering whether to use a particular collection, it is necessary to know whether there are non-adult individuals in that collection (and actually, collections solely of non-adults exist; Johns Hopkins Human Fetal Skull Collection, for instance, at the Cleveland Museum of Natural History, Cleveland, USA. See the 'Collections and Database' page on the museum's website). Aside from the basic demographic parameters, it is also useful to look at the age range for each sex individually, as the age range for one sex may be more restrictive than the other. Even where an effort has been made to include individuals from all groups in the population at large, identified collections generally underrepresent some groups.

Many collections, for example, have far more males than females – including the Grant Collection, the Maxwell Documented Collection (curated at the Maxwell Museum of Anthropology, University of New Mexico, Albuquerque, New Mexico, USA; Komar and Grivas 2008), and the Dart Collection (Tal and Tau 1983). Sex is the most important parameter to determine for skeletonised individuals in archaeological and forensic contexts, as some skeletal aging indicators have been found to vary by sex (Brooks and Suchey 1990; Gilbert and McKern 1973; Işcan *et al.* 1985). Sex is (usually) obvious and easy to estimate for individuals in anatomical and cemetery collections, as soft tissue was typically present at the time of burial or dissection. It is, however, subject to transcription errors and problems with cataloguing/storage and accidental mixing of skeletal elements.

If research is to focus on a particular part of the life course, it is also important to take absolute numbers of individuals in that part of the life course into consideration. The number of individuals at the older end of the life course is likely to vary between collections, as will the oldest age itself. For example, of the Dart, Pretoria, Spitalfields, Coimbra, Lisbon and Grant collections, only the Dart Collection lists several individuals as being over 100 years old at death, and a relatively large number of people with ages-at-death in their 80s and 90s, whereas the Grant Collection (consisting only of 202 individuals) has very few individuals with ages-at-death in their 90s (Sharman, 2014). Thus, comparing samples of oldest age individuals between collections can be difficult if the goal is to have a large enough sample size of oldest-age individuals to be statistically testable or to sample equal numbers of individuals with ages-at-death in the eighth decade or higher.

'Race' and Human Variation

The breadth of human skeletal variation sampled in the collection is another important consideration and also closely connected with age and sex. As a species, our adaptability and plasticity (the ability to react to change in the environment and activity levels), coupled with the wide range of human habitats and activities (work and leisure) that humans experience, contributes to divergent aging rates and expression of sexual dimorphism among and even within populations (e.g. Carlson *et al.* 2007; Chamberlain 2006; Falys and Lewis 2011; Kemkes-Grottenthaler 2002; Sharman 2014; Walker 2008). Humans vary by phenotype and genotype; the interaction between genes, environment, culture and behaviour is complex, and works to shape an individual's phenotype over the lifespan.

Variation, in phenotype and genes, occurs in clines, or continua, that do not coincide with 'racial' groupings (e.g. Chikhi *et al.* 1998; Handley *et al.* 2007). There is more genetic variation within any specific population than between populations, or groups referred to as a 'race' (which will hereafter appear without quotation marks; see Lewontin 1972; Brown and Armelagos 2001; Relethford 2002). Phenotypes are expressed clinally across geographic locations – there are no strict borders between races or ethnic groups (Keita and Kittles, 1997; Relethford, 2009). Race is a social-cultural construct (Keita and Kittles, 1997); biological human variation does exist, but not very much of it is centred on the socially identified racial groups. Where there are biological correlates, evidence suggests the correlation may be socio-political in origin (Gravlee, 2009) since lived experience is affected by the amount of real or perceived skin pigmentation – or other folk criteria – for grouping humans. Race was much discussed by early biological anthropologists (e.g. Hooton 1918; Hrdlička 1918), and, unfortunately, unflattering (i.e. racist) descriptions were often used for non-white, non-Western groups of people to insinuate a ranking of races (e.g. Coon 1963; see Blakey 1987 for a detailed review). While most anthropologists and about 50% of biological anthropologists (Wagner *et al.* 2017) do not agree with the concept of race as a strict biological entity now (e.g. Gravlee 2009; Relethford 2002, 2009), and certainly the racist descriptions and rankings are generally not present in current bioarchaeological and forensic anthropological research, groups of people are often still categorised in the same way as in the earlier research (e.g. Gill and Rhine 1990; Patriquin *et al.* 2003). It is highly problematic that these 19th century racial concepts of human variation that predate evolutionary theory and basic genetic concepts are still used in the 21st century. Such categorisation may disguise true (normal) variation (e.g. Stevenson *et al.* 2009). An example of racial categorisation causing a problem for interpretation of results will be presented later in this chapter.

It is important to note that we use the word race in this chapter to draw attention to the problematic nature of categorisation of humans in many of these identified collections. We are not implying that racial categories are biologically meaningful or that race is a valid biological concept. We follow the approach of Albanese and Saunders (2006), who demonstrated that the use of alternative ancestry terminology based on continental origin does not solve any of the fundamental problems inherent in a racial approach to human variation. Stating a person's 'ancestry' as 'African' is just as racial and problematic as describing a person's 'race' as 'black'. Differences in lived experience are connected to social inequality and these differences are reflected in the skeleton. Using ancestry terms implies that continental origin (not to be confused with genotype) is having the impact on the skeleton, whereas by using racial terminology in this chapter we are intentionally drawing attention to the socio-political nature of these constructed groupings and the impact of racism on the skeleton.

While typically the curator is from the general cultural group whose skeletal remains (s)he is curating (Albanese 2003b; see also Chapter 4, this volume), some concepts or ideas of group affiliation may differ between the curator and the individuals belonging to the group. For instance, in South Africa, documented ethnic affiliation for a black person may simply read 'black' or 'South Africa negro', based on identification by a white South African anatomist or doctor – however, the individual, while living, may have self-identified to a particular group (e.g. 'Xhosa' or 'Ndebele'). Alongside a 'racial' identification that the person, while alive, may not have agreed with, this practice may also result in bias in the skeletal collection – certain groups may be under- or overrepresented due to differences in the concept of group membership, and there may be a corresponding lack of detailed documentation.

Furthermore, race descriptors by third parties are affected by the political-economic context, and so may change over time. South Africa again provides an example in changing descriptors, particularly for people of colour, in the transition from apartheid to post-apartheid. In the Dart Collection, some individuals are classified as 'Mixed', 'Coloured' or 'Hybrid'; 'coloured', in South Africa, is an accepted term for a group of South Africans with mixed European, Asian and African ancestry (Dayal *et al.* 2009). People who are considered coloured also self-identify in this way, but do not seem to identify particular aspects of their mixed heritage, and would simply consider themselves South African and coloured (Billings, 2009, pers. comm.). For some individuals, tribal groups are identified (Zulu, Xhosa, and Sotho, for example, are the most common), while others are identified as SAN ('South African Negro'); skeletons of European heritage are variously called 'White', 'Euro' or 'Caucasian'. 'N/S' ('Not specified') refers to Black South Africans of unspecified population group. Dayal and colleagues (2009) note that these discrepancies were caused by South Africa's changing policies on racial classification. There is much variation between the groups represented in the Dart Collection, in terms of genetics, culture and language; furthermore, apparently even an individual in South Africa will not always consistently identify themselves as being part of a particular 'tribe' or 'race' (Dayal *et al.* 2009). This inconstancy is apparently due to a separation of the concepts of ethnicity and biology regarding identity. Further possible confusion arises from the fact that tribe was established by a third party (e.g. hospital staff) by surname or 'other contextual information' for some individuals (Dayal *et al.* 2009). However, studies on skeletal morphology have found cranial homogeneity between South African tribal groups (de Villiers 1968).

Unlike the Dart Collection, for most individuals in the Pretoria Collection, population groups are simply classified as 'White', 'Black' or 'other'. Some 'Black' individuals have more specific additional identification under 'ethnicity', including Zulu, Shangaan,

Tswana, Xhosa, although many are listed as 'Black' under both ancestry and ethnicity. Those in the 'other' ancestry category are listed as coloured under 'ethnicity' (*Kleurling* in Afrikaans), with the exception of one individual listed as Arabic (*Arabier* in Afrikaans), and a few with no specific ethnicity noted. Most white individuals are listed as such under 'ethnicity' and 'ancestry', although a few have more specific labels – Portuguese, German or Hungarian. It is not clear whether these individuals were from these other countries or if this is noted as an ethnic background.

The Terry Collection follows a pattern similar to the Pretoria Collection. Racial data are presented dichotomously, earlier in the collection period as 'White' or 'Negro' and later in the collection period as 'White' and 'Black'. While there is on-going debate about the concept of race or ancestry in physical and forensic anthropology (for example, compare the approach of Armelagos and Goodman 1998 to that of Ousley and Jantz 1998), the racial designations for the collection are highly problematic. Hunt and Albanese (2005) note several issues about race data in American anatomical collections that are also applicable to the South African collections: 1) race was designated at the time when the cadaver was used for instruction (from 1917 to 1966) and reflects the concept of race at that time; 2) the racial classification was based on social criteria and not any biological reality; 3) skin colour is not a proxy for either genetic or morphological variation; 4) the criteria for assignment to any given racial group varied as social and academic views of race changed over five decades of collecting.

Evidence from skeletons further supports point 2, above. The 'White' Dart group has diverse European ancestry, including from the Netherlands, Portugal, Germany, France and the United Kingdom (Patriquin *et al.* 2002); however, studies have suggested that the white South Africans show skeletal differences from the 'parent' populations and are distinct (e.g. Loth and Henneberg 1996; Steyn and İşcan 1998). Evidence has also shown that successive generations of children born to migrants increase in height, weight, fat and musculature to eventually match that of the host population (Bogin and Loucky 1997). When analysing variation in rates of sexual dimorphism and skeletal aging rates, Sharman (2014) found geographic clustering in terms of some skeletal traits used for sex estimation, as well as in aging rates. These clusters were at the country level – for example, the Dart and Pretoria collections had skeletal aging rates that were more similar to each other than they were to skeletal aging rates from the Lisbon and Coimbra collections (Sharman 2014). The implication is that white South Africans have more in common with black South Africans than they do with white people from other countries.

Socioeconomic Status

Typically, the socioeconomic status of individuals whose skeletal remains are curated in documented collections was low, because many rely on the skeletons of unclaimed cadavers for curation (like the Dart, Pretoria and Grant collections). However, for archaeological cemetery collections, this may not hold true. For church excavation-based collections, people buried in crypts and in lead coffins were usually wealthier (middle and upper class) than those buried in the churchyard (Litten 2002); thus, samples exclusively from crypts are not representative of the population at large, as they consist of a more privileged subset of the population. Burial location within parish churches may also signify different socioeconomic statuses, as intramural burial within the chancel was most sought-after; for those who could not afford intramural vault burial within the church, churchyard space as close as possible to the building was deemed preferable (Litten 2002). The Spitalfields named sample is particularly interesting to use as an example of the effect of socioeconomic status: firstly, because of the relative wealth of information available on occupation and associated socioeconomic status of the individuals, and secondly, because a change occurs in the average socioeconomic status of individuals over the time period the cemetery was in use.

Cox (1996; also see Molleson and Cox, 1993) undertook the collation of biographical histories for individuals belonging to the named sample. Records of baptism, marriage and burial were consulted, as were non-church records, such as trade directories, newspapers, personal papers, coroners' reports, and death certificates (Cox 1996). Information for some individuals includes details of occupation, address, socioeconomic status, and family size. Many of the families of the Spitalfields parish were of Huguenot origin; for these families particularly, living descendants were able to provide further information (Cox 1996). The Huguenots were French Protestants who left France to avoid religious persecution between the late 16th and mid-18th centuries (Cox 1996). A large number of the Huguenots who eventually settled in Spitalfields were involved in the French silk industry; these refugees brought their trade with them, making Spitalfields the hub of the silk industry in England (Cox 1996). This is reflected in the listed occupations of the named sample; of the 237 individuals for whom occupation was identified, 40% of these were within the silk industry, including master weavers, journeyman weavers and silk dyers (Cox 1996). Other occupations are diverse, ranging from high status positions such as a Member of Parliament and a surgeon, to those in construction or the food and retail industries, to low status occupations such as a bird dealer and a brush-maker (Cox 1996). Many were of the so-called 'middling sort' and artisans (of lower socio-economic status than professionals, merchants or master craftsmen); interestingly, a shift in socioeconomic status is found in Spitalfields with the turn of the 19th century. The 18th century deaths largely reflect master craftsmen, while the next century's dead were more likely to have been artisans, reflecting a change in the socioeconomic status of the population in the Spitalfields area (Cox 1996). In terms of material wealth, artisans were the only group not to own property (Cox 1996).

Despite the fact that we have some information about the socioeconomic status (the 'middling sort') of the individuals of Spitalfields Collection and, for example, of the Dart Collection, likely mostly of lower socioeconomic status, it is nevertheless difficult to make any decision about the relative socioeconomic status of a Spitalfields individual compared to a Dart individual. How does one compare the lifestyle, including occupation, quality of living space, or quality of diet of a relatively middle class 18th-century master weaver in England to that of an early 20th-century migrant worker in Johannesburg? Furthermore, even if we were to compare two individuals of low socioeconomic status from Spitalfields and Dart, 'low' is relative to their own place, their own time. If one were to take health status into consideration, then it becomes more difficult, in terms of pre- versus post-antibiotic eras, access to medical care, etc. Variation in many other cultural variables and politico-economic context render such comparison or attempts at calibration inappropriate (see Chapter 4, this volume for an example using the Terry and Coimbra Collections).

Age

How Well-Known are 'Known' Ages?

Another issue in the use of documented collections is the reliability of the 'known' ages. In cultures or areas where it is uncommon for individuals to know their own exact ages, as in modern non-counting societies, where birth years are not accurately known, or for whom chronological age has no cultural relevance, 'age heaping' is frequent (Coale and Demeny 1983; Hopkins 1966; Scheidel 1996). It has also been noted above that for some anatomical collections where many individuals' pathways into collections result from hospital deaths and bodies remaining unclaimed, hospital staff may have estimated age at death; age heaping is likely to be frequent in those cases as well. Age heaping is the tendency to report particular terminal digits in stated ages, with the corresponding evasion of other digits; thus, an abundance of ages with terminal digits of 0 and 5 would be evident (Chamberlain 2006; Scheidel 1996). This phenomenon can also be found in paleodemographic studies relying on census data, resulting in a non-representative sample population. Age heaping may occur when one individual has reported the age information for all other family members – the individual may only have estimated the family members' ages. For example, in Roman Egypt, it was noticed that 0 and 5 were favoured terminal digits, while the terminal digit 7 was particularly avoided (Scheidel 1996). The same pattern has been found on mummy labels (labels with the name of the deceased and occasionally other information on Ptolemaic and Roman Period mummies) and tombstone inscriptions. Interestingly, the number 7 was found in many magical spells and charms; Scheidel (1996) suggests that the avoidance of 7 as a terminal digit relates to its ominous magical properties. In a more recent example, the birth years of individuals of the Lisbon Collection, Portugal, were calculated from ages-at-death (which range from birth to 98 years, and dates of death are between 1880 and 1975); here, it is possible that some ages have been misreported, because literacy in Portugal was low in the first half of the 20th century (Cardoso 2005: 31). The literacy rate for Portugal for 1860 was only 12% – compare this to the literacy rate for the UK in the same year, at 69% (Tortella 1994). By 1950, Portugal's literacy rate was 56%, while the UK's literacy rate was around 99% in that year (Tortella 1994; UNESCO 1957). These examples underline the fact that age heaping may be an inherent bias as a result of the original recorded ages in documented collections, and should be tested for in modern and ancient skeletal samples where documentary evidence of age at death is present.

Another potential issue pertaining to paleodemographic census data and reported ages in skeletal collections is that of age inflation of the oldest individuals (Meindl *et al.* 1983). This may arise when individuals do not know the precise ages of elderly relatives, or it might be to honour their respected oldest family member – a boasting point of sorts. A modern example of age exaggeration is found in Vilcambamba, Ecuador (Mazess and

Mathisen 1982). This population began getting publicity for being extremely long-lived, with many people over 100 years of age. However, subsequent investigation found that age exaggeration was actually at work, with people adding years to their ages from around 60 to 70 years old – and, in fact, no individual was 100 years or older (Mazess and Mathisen 1982). If such inflations occur, the resulting interpretations of age distributions and demographic reconstructions suffer (perhaps unknowingly) from these artificially inflated distributions (subsequent effects might include incorrect calculations of life expectancy, for example). Accordingly, caution must be taken when extremely old ages are reported in known age skeletal collections.

When Did the People Die and *When* Did Collecting Occur?

During some time periods and in some places, it may be or have been more socially acceptable to include certain individuals in identified collections depending on their race, sex, socioeconomic status and/or other factors. Certainly, some people may be more likely than others to end up in an identified collection depending on these factors. As such, it is important to consider the historical context of a collection. Factors such as whether a collection is pre-, peri-, or post-antibiotic era may impact on the demographic profile of a collection; whether people died of particular diseases and if any stigma was attached to that disease may also impact whether a skeleton was curated.

Early collections by anatomists, those considered the forefathers of biological anthropology, often were curated with the intention of carrying out research on the human races, often linked with racist attitudes about the supposed hierarchical nature of the races (for further discussion, see Fabian 2010; Redman 2016).

Those who were socially disadvantaged (poor, from a marginalized group, etc.) had a higher chance of dissection by medical students, and possible subsequent curation; an example can be found in Melbourne, Australia, in the early twentieth century (Jones 2011). Richard Berry, who became chair of anatomy at the University of Melbourne in 1905, was of the opinion that those of low socioeconomic status were inherently inferior – and thus, appropriate candidates for dissection by medical students (Jones 2011). Similar evidence of socioeconomic disadvantage resulting in disproportionate chances of dissection is available from North America – over 75% of the skeletons found under the basement floor of the Medical College of Georgia from 19th century dissections were of African Americans (Blakely and Harrington, 1997).

Varying views of the human body have also affected the composition of skeletal collections over time. For instance, prior to 1832 in the UK, doctors and medical students could only dissect the bodies of executed murderers, as further punishment after death; similar legislation also existed in Canada (Ginter 2010; MacDonald 2009). Although the Anatomy Act 1832 (UK) allowed dissection of donated bodies, and was enacted to halt the 'body snatcher' trade (interestingly, the skeleton of William Burke, of the notorious Burke and Hare body snatchers, is on display at the University of Edinburgh's Anatomical Museum), the attitude that dissection for the purposes of learning anatomy was a punishment may have lingered far longer than 1832, despite the change in law (MacDonald 2009). Indeed, the unsavoury dealings in corpses led to poor public opinion of the medical schools themselves (Hutton 2013), which seem to have survived years beyond 1832 (MacDonald 2009).

Where Did the Skeletons (People) Come From?

Local People?

Socioeconomic and political contexts are important here. If a cemetery collection exists in a place that is fairly culturally homogeneous with few immigrants, then it is likely that most people are local (although checking as many documents and sources of information as possible is, of course, highly recommended). But 'local' could mean local to a particular city or town, or the wider region or state, or even the nation as a whole – so it is important to define 'local'.

The Coimbra and Lisbon collections, as already noted, consist largely of local people, buried in the local cemeteries. As well as the city, *freguesia* (parish) of death is also given for individuals (Cardoso 2006), resulting in fairly high precision in terms of location data; not only are there data on the city an individual died in, but also the part of the city in which they died. The Spitalfields Collection is also homogeneous; as discussed earlier, individuals buried at Christ Church were Huguenots, although the vast majority were born in England. Homogeneous identified collections provide advantages for some research questions, or those analysing the impact of some aspect of environment on health (for example, Cunha 1995) or can be used in conjunction with data from other sources to construct a sample that includes a wide range of human variation (for example, Albanese 2003a).

Immigrants? From How Far?

If no supplemental information is available, country of origin will likely not be known. Surnames may provide a clue, but migration has been occurring for a long time and could indicate a family who migrated generations ago. Furthermore, living conditions between the country of origin and adopted country may differ, which may impact an individual's skeleton (e.g. stature, degree of sexual dimorphism; see Bogin and Loucky 1997; Stini 1969), depending on the age an individual immigrated, whether immigration resulted in a positive or negative change in terms of quality of environment (where 'environment' encompasses social, economic and political environment), diet, etc. If an identified collection is composed largely of immigrants, particularly if they lived in their adopted country for a fairly long time before death, then this group of immigrants may be more similar to each other in some characteristics than they are to inhabitants of their respective countries of origin, as noted earlier in this chapter.

Other individuals may be migrant workers – perhaps coming from the rural context into the urban areas for work, which could be a daily or weekly journey, or individuals might stay away from home for months at a time. Depending on the type of research question being asked, variation in skeletal traits of immigrants compared to those of the other local individuals in a collection may be important. While geographic clustering (in collections located in the same country) in some skeletal traits that are expressive of sexual dimorphism and in aging rates have been found (Sharman 2014), suggesting that including short-distance (from the same country) migrants in a local population may be appropriate, it must be cautioned that this is not true of every skeletal trait. For instance, in assessing the skull for traits scored to estimate sex, the expression of sexual dimorphism in the supraorbital margin and nuchal crest was found to significantly differ between the Lisbon and Coimbra collections, though both are located in Portugal (Sharman 2014).

Why Were Particular Skeletons Added to the Collection?

In the 'What' section above, we discussed some of the reasons for collecting and the sources of skeletons. It is also essential to consider *why* particular individuals or groups of skeletons were added to a collection.

The goal of some collectors/curators was to amass a collection that included a wide range of human variation. This approach seems to be particularly evident with anatomically-derived collections where almost all available skeletons that passed through the anatomy department seem to have been have kept in an initial phase of collection. In fact, all collections, including cemetery collections, have been subject to some form of targeted collecting to address real or perceived problems with a collection sometimes decades after the collection's inception (for detailed examples of anatomical collections see Hunt and Albanese 2005 regarding the Terry Collection; and Albanese this volume regarding the Grant Collection).

The Lisbon Collection, or Luís Lopes Collection, is one of two collections of identified human skeletal remains curated at the National Museum of Natural History and Science, in Lisbon, Portugal (Cardoso 2006). The newer collection was begun in the 1980s by Luís Lopes to replace the Ferraz de Macedo Collection, which was almost completely destroyed by fire in 1978. The collection includes over 1600 individuals, who were collected during the late 1980s until 1991. The size of the collection, the relatively narrow window when collecting occurred and the inclusion of a large number of individuals from most of the major cemeteries in Lisbon means that the collection is representative of mortality in a major city on a level that is likely not matched by any other identified collection (Cardoso 2006). However, the collecting of skeletons was reinitiated in 2000, focusing mainly on non-adults and young adults to fill in gaps in the series to address a series of specific research questions about growth and development, and forensic methods (Cardoso 2006). This selective inclusion of a demographic that is often missing from identified collections (particularly anatomical collections) greatly increases the research potential of the Lisbon Collection. However, potentially misleading results are possible if a researcher sampling the collection ignores the two-stage process involved in collecting skeletons (see example 3 below).

Research Using Documented Collections: Impact and Interrelationship of 'Who, What, When, Where, Why'

In this section, we present three examples that span many of the issues raised when the five Ws are considered. The examples directly or indirectly deal with issues related to the reasons for collecting and the source of the skeletons; who was included in the collection and why; the timing of the deaths/inclusion in the collection; and the origins of the people that were included in the collections.

Example 1: Accuracy of Age-at-Death Data and Age Heaping for the Terry Collection

Although Terry was a much less prolific author than other anatomist-collectors, he devoted more time and resources to the curation of the collection and the confirmation and cross-referencing of documentary and demographic data than most collectors. Therefore, for well over 90% of the skeletons in the Terry Collection, age at death and cause of death data are more likely to be accurate and precise than information found on death certificates for that period. More importantly, the records are clearly annotated when age at death and other data could not be independently confirmed. Terry's approach and detailed documentation went a long way in addressing problems, but he did not and could not solve all the problems with age data. There is still evidence of heaping at multiples of 5 years even after the unverified ages are left out (see figure 5 from Hunt and Albanese 2005), but this type of heaping is a persistent issue and is still a problem in census data in the 21st century. For example, there are significant discrepancies between surveys and censuses for the number of people 90 years or older in the United States from 2006 to 2008. As He and Muenchrath (2011: 21) note based on data from 2006–2008, there is a digit preference for numbers ending in 0 or 5,

> '...due to a variety of factors, including a gross ignorance of the true age, lack of birth records which makes it difficult to confirm or disconfirm a reported age, reliance by some oldest people on the knowledge of others for their own age...'

In contrast to Terry, Todd was a prolific author who wrote comprehensive and foundational papers particularly in the area of age estimation, which are still relevant after over 90 years. However, just over 25% of the 3300 individuals in the Hamman-Todd Collection have confirmed ages at death. It is ironic that the collection that was used most to develop the foundation of all modern gross morphological age estimation methods has such considerable, but not insurmountable, issues with age at death data (Lovejoy *et al.* 1985, Meindl *et al.* 1990; Todd 1920; 1921; Todd and Lyon 1924; 1925a; 1925b; 1925c).

Despite the differences in size of the Terry (1700+) versus Hamann-Todd Collections (3300+), there is a larger pool of individuals with verified ages in the Terry Collection than in the Hamann-Todd Collection. Those individuals in the Terry collection whose age is not verified should be excluded from any research involving the aging process and methods for estimating age at death. There is no way of knowing by how much the documented estimated age is different from the actual age at death. For those individuals who collectively result in heaping at certain ages, it is impossible to separate those individuals who were actually, for example, 60 years old at the time of death, versus those who may have actually been 57, 58, or 59 years and were rounded up to 60 years. However, Terry's verification process ensures that there was some independent source of information to support a given age even if some minor heaping was involved. Since the actual manifestations of the aging process on the skeleton are not noticeably different for someone who is 58 or 59 versus someone who is 60 or 61, the impact of minor variations in age is minimal or zero.

Researchers should exercise caution when dealing with collections where there is no evidence of a thorough and rigorous systematic assessment of age data for an entire collection. Simply because it was recorded, even on official documents, does not mean the age is correct. In many jurisdictions (including Missouri for the Terry Collection and Ontario for the Grant Collection), age was noted as part of identity and tracking on death certificates and anatomy registers, which had to be kept in perpetuity, by law, in case the remains were claimed at various points after reaching the anatomy department. For individuals in both collections, there are many examples of discrepancies when comparing the documented age curated with the skeletal remains versus the age documented on death certificates and/or anatomy registers. By law, the official documents could not be changed, but the documents curated with each skeleton were updated as ages and other data were verified. When dealing with collections with no evidence of independent verification of ages, one strategy is to avoid ages where heaping occurs or is expected to occur.

Obviously, those individuals whose age is certainly incorrect or even suspected to be incorrect should be excluded from any research on aging or methods for estimating age at death. However, the accuracy of age data should also be considering when dealing with other research questions. Age data should be considered when investigating patterns of sexual dimorphism (Walker 1995), the development of sex estimation methods (Albanese 2003a), and the assessment of patterns of human variation (Albanese and Saunders 2006). If mortality bias is not considered and controlled through the use of accurate age at death data, it may have an enormous impact on research even if age at death is not the focus of that research (see example 3 below).

Example 2: Race, Apartheid and the Accuracy of Age at Death in the Dart and Pretoria Collections

Historical and political context can affect accuracy of some but not necessarily all data equally; the accuracy of ages-at-death in the Dart and Pretoria collections provide a good example of this.

In the Dart Collection, the majority of South African Whites (SA Whites) have ages-at-death of 60 to 89 years, while for SA Africans, the majority of skeletons are from 30 to 59 years at death (Dayal *et al.* 2009). The age/race groups of the Pretoria Collection are similar: younger black individuals, and older white individuals. This makes possible 'ethnic' differences impossible to separate from age-related differences (see example 3 below). The younger black individuals, as noted earlier, were usually unclaimed remains, or transient migrant workers from outside Johannesburg (for the Dart Collection) with no local family to claim bodies (Dayal *et al.* 2009; Tobias 1988). Hospital staff sometimes had to estimate age for the unclaimed bodies – while sex is significantly easier to report when soft tissues are present, age is more problematic, particularly in a racist (apartheid-era) country where many of the dead hospital patients were black and most of the doctors white. Indeed, Tal and Tau (1983: 217) note that up to 25% of the recorded ages differed by over 10 years compared to estimates 'by attrition'. Thus, some of the 'known' ages represent guesses by doctors in hospital immediately before death or when completing a death certificate. Conversely, the older white individuals in the Dart and Pretoria collections often donated their bodies, for complex socioeconomic reasons. Social views of human bodies have changed, and while dissection is no longer viewed as 'punishment for being poor', avoidance of funeral costs, possibly associated with low socioeconomic status is still a reason for some donations of bodies for medical school dissection and anatomical identified collections. Though Tobias (1988) suggested that the donated cadavers more likely represented individuals of higher socioeconomic status, Patriquin *et al.* (2002) note that while black South Africans were more likely to have been unclaimed and white South Africans were more likely to have been donated to the Dart Collection (and the Pretoria Collection), the donations were often precipitated by inability to pay for burial. Whatever the individual pathway into the Dart and Pretoria collections, no specific mention of age estimation of white individuals has been seen in the literature; the donation pathway is better documented. Despite no discussion of age estimates of white individuals, the possibility should still be tested for – for all individuals, regardless of stated ethnicity. Evidence for age estimates would come by way of a high frequency of ages with terminal digits of 5 and 0 – the aforementioned age heaping, as when people guess the ages of others, they tend to round to the nearest five years.

Age heaping was tested for by constructing distribution frequencies of ages-at-death for each collection (Sharman 2014). As the youngest age at death used for skeletons in this research was 20 years old, this was the youngest included age in the frequency distributions. The oldest included age was 114 years, an individual from the Dart Collection, and this was the oldest reported age out of all of the collections, suspicions of the possibility of age exaggeration notwithstanding.

Age heaping was found in both the Dart and Pretoria Collections. The Dart results were not surprising, given the previously published results of Tal and Tau (1983) and Dayal and colleagues (2009); Dayal and colleagues (2009) suggest that while many ages ending in 0 or 5 are estimates, ages in between these intervals are likely accurate. Age heaping was found for all ethnicities and both sexes, although it was most pronounced in non-white ethnicities. Interestingly, the age heaping found in documented ages from the Pretoria Collection seem to be confined to individuals with ethnicities listed as black or 'other' ('other' refers to 'coloured' individuals in the Pretoria Collection and individuals listed as 'hybrid' or 'mixed' in the Dart Collection). White individuals do not show evidence of age heaping. These differences are not constrained by sex; males and females listed as black or 'other' exhibit age heaping, while males and females listed as white do not. It was also possible to divide the Pretoria Collection sample into during- and post-apartheid years, to see whether the racist attitudes of apartheid-era South Africa affected the assessment of age of individuals dying in hospital by staff. However, no differences were noted when the collection was divided by those who died before 1994 and those who died from 1994 to the present.

Variation in individuals' pathways into the collection is also a probable factor in the racial division in age heaping. Better documentation is likely available for individuals who willed their bodies or whose families donated their bodies to the collection; if the division between willed/donated and unclaimed bodies follows racial lines, age heaping could be due to the fact that younger, black, 'other', or 'coloured' individuals were more likely to be migrant workers, or of low socioeconomic class with families unable to afford burial. L'Abbe and colleagues (2005) note that black South Africans tend not to donate their bodies or bodies of relatives due to ancestor reverence. Clearly, this is at odds with the high prevalence of black males in both the Dart and Pretoria Collections. However, when young black males migrate to cities from rural areas to find work, and then die in the city, it is highly problematic for hospital workers to find and contact immediate family of the deceased (L'Abbe *et al.* 2005); their bodies remain unclaimed and hospital staff then must estimate their age. Conversely, black females in South Africa typically do not leave their home areas for work, and thus their bodies are more likely to be claimed upon death, explaining the higher ratio of young black males to young black females. Older people are less likely to be migrant workers, and thus, probably lived fairly locally, which perhaps means these people are more likely to have family nearby who could provide accurate age information for the deceased. Data also show that, in 1990, over 80% of older black and other non-white South Africans lived with two, three or more generations, while 80% of older white South Africans lived alone or with only their spouse (Kinsella and Ferreira 1997), perhaps adding to the chances of older black South Africans being claimed upon death compared to older white South Africans. It would be very interesting to examine individual pathways into collections and race – if white people are more likely to have willed or donated their bodies, while black, 'other', or 'coloured' people are more likely to be unclaimed upon death in hospital, then pathway into the collection would be the major influence in age at death estimation, and thus, accuracy or inaccuracy of age at death. However, neither the Dart nor Pretoria collection data lists include information about pathway into the collection (although it is possible that more extensive records are kept by curators; this information was not requested at the time the collections were accessed for research. Future researchers may wish to ask that question).

The conclusion here is that variables other than race are much better for explaining the patterns of variation: age at death (younger people who migrated for work and died in hospital), socioeconomic factors, poverty and differential access to resources (e.g. jobs local to where individuals live).

Example 3: Misinterpretation of Mortality Bias in the Terry Collection as 'Racial Differences'

In many reference collections, the relatively narrow window of collecting means that year of birth (YOB) is correlated with age at death. The general trend is that the youngest adults who may manifest a mortality bias (morbidity and poverty during growth and development resulting in early mortality) have the most recent years of birth and should show the greatest effects of a positive secular trend in adult stature. When a collection is larger and the collection period is longer, as is the case with the Terry Collection, it is possible to sample the collection to control and/or assess the combined impact of both of these variables and investigate patterns of variation (Albanese 2003a). However, the historical context of the collecting process must also be considered. With the Terry Collection, there was a significant shift in the collection process after Terry retired in 1941 and Mildred Trotter attempted to balance the race and sex ratios in the collection and selectively added mostly older white females to the collection (see Chapter 4, this volume).

Studies that do not consider the impact of both mortality bias and secular changes within the context of the history of the collection process will necessarily produce misleading results. This potential problem can be illustrated using an example for assessing patterns of variation in the pubic bone in a sample from the Terry Collection (see Albanese and Saunders 2006 for a detailed presentation of the statistical analysis). The sample for analysis was selected to minimize the impact of YOB *and* age effects as much as possible using a multistep stratified sampling approach. The sub-samples could be matched for YOB and age for black and white males, but because of Trotter's influence, the sample could be matched for YOB but not age for black and white females.

When comparing the mean pubic bone length for these sub-samples from the Terry Collection, there was no significant difference in pubic bone length between black males and white males, but white females *seem* to have significantly longer pubic bones than black females. The results seem to be consistent with previous research suggesting that after the skull, the pelvis is a good source of information for estimating ancestry or race (DiBennardo and Taylor 1983; İşcan 1981, 1983; İşcan and Cotton 1984; Krogman and İşcan 1986; Schulter-Ellis and Hayek 1984). However, when age data are considered in the historical context of Trotter's impact on the collection, another more likely explanation becomes obvious: pubic bone length is significantly correlated with age and a comparison of black females to white females in the Terry Collection is actually a comparison of younger adult females to older adult females. The results were similar for other highly sexually dimorphic dimensions of the pelvis, including iliac breadth.

While all the YOB and age bias could not be controlled for, it could be clearly identified: race is correlated with age at death due to the collection process. Those females with compromised pelvic growth and development, who coincidently happen to have been described as 'Black/Negro' using 19th century concepts of human variation, died younger. Cause-of-death data are available and none of these females died in child birth. Rather, the results show that compromised pubic bone growth is a non-specific stress indicator related to premature mortality in females. If anything, the significant differences are due to impacts of racism and economic disparity on growth and development, and not race.

This example shows that not considering these biases will have two major impacts on research. First, any methods for estimating race or ancestry (and any race-specific sex estimation methods as well) that are based on mortality bias rather than group-specific differences will provide misleading results when applied outside of the reference collection in actual forensic contexts or bioarchaeological applications. Second, a racial or typological approach to research where skeletal elements are considered without any context will result in a poorer understanding of actual human skeletal variation.

Conclusion

There are several types of documented collections that can be grouped by source (cemetery versus anatomical) or by the process by which they were amassed (*post hoc* versus constructed). The value of these collections is directly derived from the documentary data that are available for each individual and the collection as a whole. However, the reasons for collecting and the source of skeletons has had an enormous influence on who is included in the collection and what documentary data are available for research, and thus the demographic composition of collections, the range of skeletal variation in collections and the types of questions that can be addressed using those collections. For example, a method for estimating age derived only from a sample from Spitalfields would be of limited utility when applied outside of Spitalfields – and indeed, Buckberry and Chamberlain's 2002 auricular surface method used Spitalfields as a reference collection, and the expressions of surface texture and transverse organization necessary for the highest scores were very rare in the Dart and Pretoria collections, which has implications for the method's ability to estimate old age in the Dart and Pretoria collections (Sharman, 2014). However, this collection would be ideal for testing existing methods to better understand their performance (e.g. accuracy or level of bias) when applied to a mortuary sample with no documentary data.

In this chapter, we have focused on issues related to age and sex for several reasons. First, age and sex are the foundation of much, if not all, research using skeletal data. For example: how does normal and pathological skeletal variation vary by sex? How do sex and gender impact on normal variation? How does age at death impact on skeletal variation? How do these patterns vary through time and space in various archaeological contexts? Second, age and sex may be major sources of bias in the patterns of human variation seen in skeletal collections, but their impact can also be relatively easy to assess and quantify, and can be used as proxies for revealing or controlling other biases in the reference collections themselves and other undocumented collections. Furthermore, documented collections are important tools for conducting research on past populations. Only by using documented collections is it possible to develop, test and continuously refine sex and age estimation methods that can then be applied in bioarchaeological contexts to address anthropological questions about past populations.

Much of the criticism of documented collections has focused on the age of the collections. While older documented collections can be problematic, depending on the research question, older collections may be particularly useful, specifically because of their age, if context is considered. For example, developing a protocol for differential diagnosis of a specific infectious disease in past populations is only possible with a sample from a pre-antibiotic collection. All collections, whether documented or not, are unrepresentative in some aspect of the mortality pattern and the living population from which they were drawn. Newer collections provide additional sources of data but are not necessarily 'better' than older collections. No collection, regardless of its age, is an 'osteological census' of human variation.

Many researchers do not spend enough time evaluating the quality and accuracy of the demographic data and the historical context of collections. The documented collections are only as good as their documents. Errors may occur in specific data for a given individual, and systemic errors or bias related to the political, economic and historical context in which collecting occurred (or is still occurring) are likely. As evidenced by the examples described in this chapter, all documentary data must be critically assessed on various levels and checked for errors. Research time is always finite and a simple step such as creating a bar graph of age by sex and ethnic group before sample selection can take time away from collecting skeletal data, but is essential. There are specific issues for each collection and research question that need to be considered when selecting samples and different frameworks or approaches arriving at the same end. In this chapter, we presented a framework comprising the 5 Ws, alongside background information and various examples, which can usefully be applied to identify potential biases or gaps in any collection. We also discussed two Hs: **how** did individuals or groups come to be part of a collection and **how** did collectors/curators verify the documentary data? Considering these questions when conducting research on documented collections is the difference between *describing apparent patterns* of human skeletal variation versus *understanding and explaining* the complex interactions of factors on individuals and groups that manifest as patterns of variation.

References

Adams, M. and Reeve, J. 1987. Excavations at Christ Church, Spitalfields 1984–6. *Antiquity* 61: 247–256.

Albanese, J. 2003a. A metric method for sex determination using the hipbone and the femur. *Journal of Forensic Sciences* 48: 1–11.

Albanese, J. 2003b. *Identified Skeletal Reference Collections and the Study of Human Variation.* Unpublished PhD thesis, McMaster University.

Albanese, J. 2013. A method for determining sex using the clavicle, humerus, radius and ulna. *Journal of Forensic Science* 58: 1413–1419.

Albanese, J. and Saunders, S. R. 2006. Is it possible to escape racial typology in forensic identification? In A. Schmitt, E. Cunha and J. Pinheiro (eds), *Forensic Anthropology and Medicine: Complementary Sciences From Recovery to Cause of Death*: 281–315. Totowa, Humana Press.

Albanese, J., Tuck, A., Gomes, J. and Cardoso, H. F. V. 2016. An alternative approach for estimating stature from long bones that is not population- or group-specific. *Forensic Science International* 259: 59–68.

Armelagos, G. J., Carlson, D. S. and Van Gerven, D. P. 1982. The theoretical foundations and development of skeletal biology. In F. Spencer (ed.), *A History of American Physical Anthropology 1930-1980*: 305–328. New York, Academic Press.

Armelagos, G. J. and Goodman, A. H. 1998. Race, racism, and anthropology. In A. H. Goodman and T. L. Leatherman (eds), *Building a New Biocultural Synthesis: Political-Economic Perspectives on Human Biology*: 359–378. Ann Arbor, University of Michigan Press.

Armelagos, G. J. and Van Gerven, D. P. 2003. A century of skeletal biology and paleopathology: contrasts, contradictions, and conflicts. *American Anthropologist* 105: 53–64.

Bedford, M. E., Russell, K. F., Lovejoy, C. O., Meindl, R. S., Simpson, S. W. and Stuart-Macadam, P. L. 1993. Test of multifactorial aging method using skeletons with known ages-at-death from the Grant Collection. *American Journal of Physical Anthropology* 91: 287–297.

Bello, S. M., Thomann, A., Signoli, M., Dutour, O. and Andrews, P. 2006. Age and sex bias in the reconstruction of past population structures. *American Journal of Physical Anthropology* 129: 24–38.

Blakely, R. I. and Harrington, J. M. 1997. *Bones in the Basement: Postmortem Racism in Nineteenth-Century Medical Training*. Washington, Smithsonian Institution Scholarly Press.

Blakey, M. L. 1987. Skull doctors: intrinsic social and political bias in the history of American physical anthropology, with special reference to the work of Ales Hrdlicka. *Critique of Anthropology* 7: 7–35.

Bogin, B. and Loucky, J. 1997. Plasticity, political economy, and physical growth status of Guatemala Maya children living in the United States. *American Journal of Physical Anthropology* 102: 17–32.

Brooks, S. T. and Suchey, J. M. 1990. Skeletal age determination based on the os pubis: a comparison of the Acsadi-Nemeskeri and Suchey-Brooks method. *Human Evolution* 5: 227–238.

Brown, R. A. and Armelagos, G. J. 2001. Apportionment of racial diversity: a review. *Evolutionary Anthropology: Issues, News and Reviews* 10: 34–40.

Buckberry, J. L. and Chamberlain, A. T. 2002. Age estimation from the auricular surface of the ilium: a revised method. *American Journal of Physical Anthropology* 119: 231–239.

Buikstra, J. E. and Gordon, C. C. 1981. The study and restudy of human skeletal series: the importance of long-term curation. *Annals of the New York Academy of Sciences* 376: 449–465.

Buikstra, J. E. and Konigsberg, L. W. 1985. Paleodemography: critiques and controversies. *American Anthropologist* 87: 316–333.

Buikstra, J. E. and Ubelaker, D. H. (eds.) 1994. *Standards for Data Collection from Human Skeletal Remains*. Fayetteville, Arkansas Archaeological Survey.

Cardoso, H. F. V. 2005. *Patterns of Growth and Development of the Human Skeleton and Dentition in Relation to Environmental Quality: A Biocultural Analysis of a Sample of 20 Century Portuguese Subadult Documented Skeletons*. Unpublished PhD thesis, McMaster University.

Cardoso, H. F. V. 2006. Brief communication: the collection of identified human skeletons housed at Bocage Museum (National Museum of Natural History), Lisbon, Portugal. *American Journal of Physical Anthropology* 129: 172–176.

Cardoso, H. F. V., Marinho, L. and Albanese, J. 2016. Relationship between cadaver, living and forensic stature: a review of current knowledge and a test using a sample of adult Portuguese males. *Forensic Science International* 258: 55–63.

Carlson, K. J., Grine, F. E. and Pearson, O. M. 2007. Robusticity and sexual dimorphism in the postcranium of modern hunter-gatherers from Australia. *American Journal of Physical Anthropology* 134: 9–23.

Caspari, R. 2009. 1918: three perspectives on race and human variation. *American Journal of Physical Anthropology* 139: 5–15.

Chamberlain, A. 2006. *Demography in Archaeology.* Cambridge, Cambridge University Press.

Chikhi, L., Destro-Bisol, G., Pascall, V., Baravelli, V., Dobosz, M. and Barbujani, G. 1998. Clinal variation in the nuclear DNA of Europeans. *Human Biology* 70: 643–657.

Cleveland Museum of Natural History website. *Collections and Database* page. Accessed online 25/02/2017 https://www.cmnh.org/phys-anthro/collection-database

Coale, A. J. and Demeny, P. 1983. *Regional Model Life Tables and Stable Populations.* Toronto, Academic Press.

Coon, C. C. 1963. *The Origin of Races.* Knopf, New York.

Coqueugniot, H. and Weaver, T. D. 2007. Brief communication: infracranial maturation in the skeletal collection from Coimbra, Portugal: new aging standards for epiphyseal union. *American Journal of Physical Anthropology* 134: 424–437.

Cox, M. 1996. *Life and Death in Spitalfields, 1700 to 1850.* York, Council for British Archaeology.

Cunha, E. 1995. Testing identification records: evidence from the Coimbra Identified Skeletal Collections (nineteenth and twentieth centuries). In S. R. Saunders and A. Herring (eds), *Grave Reflections: Portraying the Past Through Cemetery Studies*: 179–198. Toronto, Canadian Scholars Press.

Dayal, M. R., Kegley, A. D. T., Strkalj, G., Bidmos, M. A. and Kuykendall, K. L. 2009. The history and composition of the Raymond A. Dart collection of human skeletons at the University of the Witwatersrand, Johannesburg, South Africa. *American Journal of Physical Anthropology* 140: 324–335.

De Villiers, H. 1968. Sexual dimorphism of the South African Bantu-speaking negro. *South African Journal of Science* 64: 118–124.

Dias, N. 1989. Séries de cranes et armée de squelettes : les collections anthropologiques en France dans la seconde moitié du XIXe siecle. *Bulletins et Mémoires de la Société d'Anthropologie de Paris* 1: 203–230.

DiBennardo, R. and Taylor, J. V. 1983. Multiple discriminant function analysis of sex and race in the postcranial skeleton. *American Journal of Physical Anthropology* 61: 305–314.

Fabian, A. 2010. *The Skull Collectors: Race, Science, and America's Unburied Dead.* Chicago, The University of Chicago Press.

Falys, C. G. and Lewis, M. E. 2011. Proposing a way forward: a review of standardisation in the use of age categories and ageing techniques in osteological analysis (2004-2009). *International Journal of Osteoarchaeology* 21: 704–716.

Ferreira, M. T., Vicente, R., Navega, D., Gonçalves, D., Curate, F. and Cunha, E. 2014. A new forensic collection housed at the University of Coimbra, Portugal: the 21 century identified skeletal collection. *Forensic Science International* 245: 202.e1–202.e5.

Gilbert, B. M. and McKern, T. W. 1973. A method for aging the female os pubis. *American Journal of Physical Anthropology* 38: 31–38.

Gill, G. W. and Rhine, S. (eds.) 1990. *Skeletal Attribution of Race: Methods for Forensic Anthropology.* Anthropological Papers, No. 4. Albuquerque, Maxwell Museum of Anthropology.

Ginter, J. K. 2010. Origins of the odd fellows skeletal collection: exploring links to early medical training. In C. Ellis, N. Ferris, P. Timmins and C. White (eds), *The 'Compleat Archaeologist': Papers in Honor of Michael W. Spence*: 187–203. Ontario Archaeology No. 85–88. London Chapter, OAS Occasional Publication No. 9.

Gravlee, C. C. 2009. How race becomes biology: embodiment of social inequality. *American Journal of Physical Anthropology* 139: 47–57.

Handley, L. J. L., Manica, A., Goudet, J. and Balloux, F. 2007. Going the distance: human population genetics in a clinal world. *Trends in Genetics* 23: 432–439.

He, W. and Muenchrath, M. 2011. US Census Bureau, American Community Survey Reports, ACS-17, 90+ in the United States: 2006–2008. Washington, DC, US Government Printing Office.

Henderson, C. Y., Craps, D. D., Caffell, A. C., Millard, A. R. and Gowland, R. 2013. Occupational mobility in 19 century rural England: the interpretation of entheseal changes. *International Journal of Osteoarchaeology* 23: 197–210.

Hooton, E. A. 1918. On certain Eskimoid characters in Icelandic skulls. *American Journal of Physical Anthropology* 1: 53–76.

Hopkins, K. 1966. On the probable age structure of the Roman population. *Population Studies* 20: 245–264.

Hoppa, R. D. 1999. Modeling the effects of selection-bias on palaeodemographic analyses. *Homo* 50: 228–243.

Hrdlička, A. 1918. Physical anthropology: its scope and aims; its history and present status in America. *American Journal of Physical Anthropology* 1: 3–23.

Hunt, D. R. and Albanese, J. 2005. History and demographic composition of the Robert J. Terry Anatomical Collection. *American Journal of Physical Anthropology* 127: 406–417.

Hutton, F. 2013. *The Study of Anatomy in Britain, 1700-1900*. The Body, Gender and Culture, Number 13. New York, Routledge.

İşcan, M. Y. 1981. Race determination from the pelvis. *Ossa* 8: 95–100.

İşcan, M. Y. 1983. Assessment of race from the pelvis. *American Journal of Physical Anthropology* 62: 205–208.

İşcan, M. Y. and Cotton, T. S. 1984. The effects of age on the determination of race. *Collegium Antropologicum* 8: 131–138.

İşcan, M. Y., Loth, S. R. and Wright, R. K. 1985. Age estimation from the rib by phase analysis: white females. *Journal of Forensic Sciences* 30: 853–863.

Jones, R. L. 2011. Cadavers and the social dimension of dissection. In S. Ferber and S. Wilde (eds), *The Body Divided: Human Beings and Human 'Material' in Modern Medical History*: 29–52. Burlington, Ashgate Publishing Company.

Keita, S. O. Y. and Kittles, R. A. 1997. The persistence of racial thinking and the myth of racial divergence. *American Anthropologist* 99: 534–544.

Kemkes-Grottenthaler, A. 2002. Aging through the ages: historical perspectives on age indicator methods. In R. D. Hoppa and J. W. Vaupel (eds), *Paleodemography - Age Distributions from Skeletal Samples*: 48–72. Cambridge, Cambridge University Press.

Kinsella, K. and Ferreira, M. 1997. International Brief. Aging Trends: South Africa. US Department of Commerce, Economic and Statistics Administration, Bureau of the Census. Available at: https://www.census.gov/population/international/files/ib-9702.pdf>

Komar, D. A. and Grivas, C. 2008. Manufactured populations: what do contemporary reference skeletal collections represent? A comparative study using the Maxwell Museum Documented Collection. *American Journal of Physical Anthropology* 137: 224–233.

Krogman, W. and İşcan, M. Y. 1986. *The Human Skeleton in Forensic Medicine*. Springfield, Charles C. Thomas.

L'Abbe, E. N., Loots, M. and Meiring, J. H. 2005. The Pretoria bone collection: a modern South African skeletal sample. *HOMO – Journal of Comparative Human Biology* 56: 197–205.

Lewontin, R. C. 1972. The apportionment of human diversity. *Evolutionary Biology* 6: 381–398.

Litten, J. 2002. *The English Way of Death: The Common Funeral Since 1450.* London, Robert Hale.

Loth, S. R. and Henneberg, M. 1996. Mandibular ramus flexure: a new morphologic indicator of sexual dimorphism in the human skeleton. *American Journal of Physical Anthropology* 99: 473–485.

Lovejoy, C. O., Meindl, R. S., Mensforth, R. P. and Barton, T. J. 1985. Multifactorial determination of skeletal age at death: a method and blind test of its accuracy. *American Journal of Physical Anthropology* 68: 1–14.

MacDonald, H. 2009. Procuring corpses: the English Anatomy Inspectorate, 1842 to 1858. *Medical History* 53: 379–396.

Mazess, R. B. and Mathisen, R. W. 1982. Lack of unusual longevity in Vilcabamba, Ecuador. *Human Biology* 54: 517–524.

Meindl, R. S., Lovejoy, C. O. and Mensforth, R. P. 1983. Skeletal age-at-death: accuracy of determination and implications for human demography. *Human Biology* 55: 73–87.

Meindl, R. S., Russell, K. F. and Lovejoy, C. O. 1990. Reliability of age-at-death in the Hamann-Todd Collection: validity of subselection procedures used in blind tests of the summary age technique. *American Journal of Physical Anthropology* 83: 349–357.

Mitchell, P. D. (ed.) 2012. *Anatomical Dissection in Enlightenment Britain and Beyond: Autopsy, Pathology and Display*. Aldershot, Ashgate.

Molleson, T. and Cox, M. 1993. *The Spitalfields Project. Volume 2 – The Anthropology. The Middling Sort.* CBA research report 86. York, Council for British Archaeology.

Native American Graves Protection and Repatriation Act (NAGPRA). 1990. *Native American Graves Protection and Repatriation Act* (25 US Code 3001 et seq.) Washington, DC, National Park Service, US Department of the Interior. Accessed online 26/02/2017 <https://www.nps.gov/nagpra/mandates/25usc3001etseq.htm>

Ousley, S. and Jantz, R. L. 1998. The forensic data bank: documenting skeletal trends in the United States. In K. Reichs (ed.), *Forensic Osteology*: 297–315. Springfield, Charles C. Thomas.

Patriquin, M. L., Loth, S. R. and Steyn, M. 2003. Sexually dimorphic pelvic morphology in South African whites and blacks. *HOMO – Journal of Comparative Human Biology* 53: 255–262.

Patriquin, M. L., Steyn, M. and Loth, S. R. 2002. Metric assessment of race from the pelvis in South Africans. *Forensic Science International* 127: 104–113.

Redman, S. J. 2016. *Bone Rooms: From Scientific Racism to Human Prehistory in Museums.* Cambridge, Harvard University Press.

Relethford, J. H. 2002. Apportionment of global human genetic diversity based on craniometrics and skin color. *American Journal of Physical Anthropology* 118: 393–398.

Relethford, J. H. 2009. Race and global patterns of phenotypic variation. *American Journal of Physical Anthropology* 139: 16–22.

Rocha, A. 1995. Les collections ostéologiques humaines identifiées du Musée Anthropologigue de l'Université de Coimbra. *Antropologia Portuguesa* 13: 7–38.

Rühli, F. J., Hotz, G. and Böni, T. 2003. Brief communication: the Galler Collection. A little-known historic Swiss bone pathology reference series. *American Journal of Physical Anthropology* 121: 15–18.

Santos, A. L. 2000. *A Skeletal Picture of Tuberculosis*: Macroscopic, Radiological, Biomolecular and Historical Evidence from the Coimbra Identified Skeletal Collection. Unpublished PhD thesis, University of Coimbra.

Santos, A. L. and Roberts, C. A. 2001. A picture of tuberculosis in young Portuguese people in the early 20 century: a multidisciplinary study of the skeletal and historical evidence. *American Journal of Physical Anthropology* 115: 38–49.

Santos, A. L. and Roberts, C. A. 2006. Anatomy of a serial killer: differential diagnosis of tuberculosis based on rib lesions of adult individuals from the Coimbra Identified Skeletal Collection, Portugal. *American Journal of Physical Anthropology* 130: 38–49.

Saunders, S. R. and Herring, A. (eds) 1995. *Grave Reflections: Portraying the Past Through Cemetery Studies*. Toronto, Canadian Scholars Press.

Saunders, S. R., Herring, A., Sawchuk, L. A. and Boyce, G. 1995. The nineteenth century cemetery at St. Thomas' Anglican Church, Belleville: skeletal remains, parish records, and censuses. In S. R. Saunders and A. Herring (eds), *Grave Reflections: Portraying the Past Through Cemetery Studies*: 93–117. Toronto, Canadian Scholars Press.

Scheidel, W. 1996. What's in an age? A comparative view of bias in the census returns of Roman Egypt. *Bulletin of the American Society of Papyrologists* 33: 25–59.

Scheidel, W. 2001a. Progress and problems in Roman demography. In W. Scheidel (ed.), *Debating Roman Demography*: 1–82. Boston, Brill.

Scheidel, W. 2001b. Roman age structure: evidence and models. *The Journal of Roman Studies* 91: 1–26.

Schulter-Ellis, F. P. and Hayek, L. A. 1984. Predicting race and sex with an acetabulum/pubis index. *Collegium Antropologicum* 8: 155–162.

Sharman, J. 2014. *Age, Sex and the Life Course: Population Variability in Human Ageing and Implications for Bioarchaeology*. Unpublished PhD thesis, Durham University.

Stevenson, J. C., Mahoney, E. R., Walker, P. L. and Everson, P. M. 2009. Technical note: prediction of sex based on five skull traits using decision analysis (CHAID). *American Journal of Physical Anthropology* 139: 434–441.

Steyn, M. and İşcan, M. Y. 1998. Sexual dimorphism in the crania and mandibles of South African whites. *Forensic Science International* 98: 9–16.

Steyn, M. and İşcan, M. Y. 1999. Osteometric variation in the humerus: sexual dimorphism in South Africans. *Forensic Science International* 106: 77–85.

Stini, W. A. 1969. Nutritional stress and growth: sex difference in adaptive response. *American Journal of Physical Anthropology* 31: 417–426.

Tal, H. and Tau, S. 1983. Statistical survey of the human skulls in the Raymond Dart Collection of skeletons. *South African Journal of Science* 79: 215–217.

Terry, R. J. 1940. On measuring and photographing the cadaver. *American Journal of Physical Anthropology* 26: 433–447.

Tobias, P. 1988. War and the negative secular trend of South African blacks with observations on the relative sensitivity of cadaveric and non-cadaveric populations to secular effects. In *Proceedings of the 5 Congress of the European Anthropological Association* I: 451–461. Lisbon.

Tobias, P. V. 1987. Memories of Robert James Terry (1871–1966) and the genesis of the Terry and Dart collections of human skeletons. *Adler Museum Bulletin* 13: 31–34.

Tobias, P. V. 1991. On the scientific, medical, dental and educational value of collections of human skeletons. *International Journal of Anthropology* 6: 277–280.

Todd, T. W. 1920. Age changes in the pubic bone: I. The male white pubis. *American Journal of Physical Anthropology* 3: 285–334.

Todd, T. W. 1921. Age changes in the pubic bone: II, the pubis of the male negro-white hybrid; III, the pubis of the white female; IV, the pubis of the female negro-white hybrid. *American Journal of Physical Anthropology* 4: 1–70.

Todd, T. W. and Lyon, D. W. 1924. I Endocranial suture closure in the adult males of white stock. *American Journal of Physical Anthropology* 7: 325–384.

Todd, T. W. and Lyon, D. W. 1925a. II Ectocranial suture closure in the adult males of white stock. *American Journal of Physical Anthropology* 8: 23–45.

Todd, T. W. and Lyon, D. W. 1925b. Endocranial suture closure in the adult males of negro stock. *American Journal of Physical Anthropology* 8: 47–71.

Todd, T. W. and Lyon, D. W. 1925c. Ectocranial suture closure in the adult males of negro stock. *American Journal of Physical Anthropology* 8: 149–168.

Tortella, G. 1994. Patterns of economic retardation and recovery in South-Western Europe in the nineteenth and twentieth centuries. *Economic History Review* 47: 1–21.

Trotter, M. 1966. Robert James Terry, MD— obituary. *American Journal of Physical Anthropology* 25: 97–98.

Trotter, M. 1981. Robert James Terry, 1871–1966. *American Journal of Physical Anthropology* 56: 503–508.

UNESCO. 1957. World Illiteracy at Mid-Century: A Statistical Study. Monographs on Fundamental Education – XI. Paris, United Nations Educational, Scientific and Cultural Organization. Available at: http://unesdoc.unesco.org/images/0000/000029/002930eo.pdf

Usher, B. M. 2002. Reference samples: the first step in linking biology and age in the human skeleton. In R. D. Hoppa and J. W. Vaupel (eds), *Paleodemography - Age Distributions from Skeletal Samples*: 29–47. Cambridge, Cambridge University Press.

Walker, P. L. 1995. Problems of preservation and sexism in sexing: some lessons from historical collections and palaeodemographers. In S. R. Saunders and A. Herring (eds), *Grave Reflections: Portraying the Past Through Cemetery Studies*: 31–47. Toronto, Canadian Scholars Press.

Walker, P. L. 2008. Sexing skulls using discriminant function analysis of visually assessed traits. *American Journal of Physical Anthropology* 136: 39–50.

Walker, P. L., Johnson, J. R. and Lambert, P. M. 1988. Age and sex biases in the preservation of human skeletal remains. *American Journal of Physical Anthropology* 76: 183–188.

Wagner, J. K., Joon-Ho Yu, J. H., Ifekwunigwe, J. O., Harrell, T. M., Bamshad, M. J. and Royal, C. D. 2017. Anthropologists' views on race, ancestry, and genetics. *American Journal of Physical Anthropology* 162: 318–327.

Wasterlain, S. N., Hillson, S. and Cunha, E. 2009. Dental caries in a Portuguese identified skeletal sample from the late 19 and early 20 centuries. *American Journal of Physical Anthropology* 140: 64–79.

Zuckerman, M. K., Kamnikar, K. and Mathena, S. 2014. Recovering the 'body politic': a relational ethics of meaning for bioarchaeology. *Cambridge Archaeological Journal* 24: 1–9.

Chapter 6

The significance of identified human skeletal collections to further our understanding of the skeletal ageing process in adults

Vanessa Campanacho[1,2,3] and Hugo F.V. Cardoso[4]

[1] LABOH - Laboratory of Biological Anthropology and Human Osteology, CRIA, New University of Lisbon

[2] CIAS - Research Centre for Anthropology and Health, University of Coimbra

[3] CEF - Centre for Functional Ecology - Group of Paleoecology and Forensic Sciences, University of Coimbra

[4] Department of Archaeology and Centre for Forensic Research, Simon Fraser University

Introduction

Identified human skeletal collections can be derived from remains collected from historical and recent cemeteries, from unclaimed cadavers used in anatomy classes or undergoing medico-legal autopsy, and from body donations. These collections are of paramount importance due to the known biographical information associated with each skeleton, combined with a greater completeness and preservation than most archaeological and forensic non-identified skeletal remains (Usher 2002). These biographical data include the sex, the age at the time of death, the last known occupation, the cause of death, and sometimes additional information such as ascribed ancestry or anthropometric data, such as cadaveric stature and weight (Cardoso 2006; Hunt and Albanese 2005; Rissech and Steadman 2011). Biographical data are typically derived from obituary records but can also be obtained from coffin plaque inscriptions (Cardoso 2006; Cunha and Wasterlain 2007; Molleson *et al.* 1993).

An identified skeletal collection represents an osteological profile of a subset of the original population, from which it was drawn, spanning a certain time period, typically from the 19th century onwards. Therefore, in addition to the biographical data, the historical, socioeconomic, geographic and cultural contexts are also known in great detail, providing researchers with the means to use a biocultural approach in their analyses of the skeletons, such as the investigation of behaviour patterns according to occupational and social categories (Alves-Cardoso 2008; Boyd 1996; Holliman 2011; Lessa 2011; Zuckerman and Armelagos 2011). Identified skeletal collections have had a very important role in the development of physical anthropology as a discipline (Eliopoulos *et al.* 2007). They are of paramount importance for research because: 1) the majority of age and sex estimation methods were established and tested using identified skeletal collections (Eliopoulos *et al.* 2007; Usher 2002); 2) these osteological methods are then applied to non-identified skeletal remains to establish a biological profile; and 3) they provide the source material for understanding population variation in size and shape, including sex- and age-related skeletal manifestations. Additionally, identified skeletal collections also have an important role in teaching (Eliopoulos *et al.* 2007; Rissech and Steadman 2011). Learning osteological concepts and methods with identified skeletal collections allow students to better understand the expressions of sexual dimorphism, pathological variation and ageing manifestations in the skeleton than they would with archaeological skeletal remains by knowing the actual sex, and age of the individual. Since age estimation research carried out by physical anthropologists is intrinsically linked to the analysis of identified skeletal collections, the present chapter reviews the research investigating the effects of environmental factors on skeletal ageing, and discusses the importance and limitations of identified skeletal collections in understanding the skeletal ageing process, as well the practical implications of this knowledge. Thus, the current chapter is divided into three main sections. Firstly, the importance of identified skeletal collections in documenting and understanding the ageing manifestations in the skeleton will be briefly discussed. In the second section, a detailed review of the literature focusing on how environmental and genetic factors may influence the expression of age-related criteria in the skeleton will be undertaken. Lastly, the limitations of studying identified skeletal collections for the purposes of age estimation will be highlighted.

The importance of understanding the ageing process with identified skeletal remains

Due to their documented ages, identified skeletal collections are a vital source of information to understand age-related bone degeneration processes. In adults, skeletal tissue undergoes degenerative modifications over time, which can be quantified and used to develop age estimation techniques[1] based on the association between the actual age of the individual and the changes in the skeletal indicator of degeneration. For adults, the most used age estimation methods involve macroscopic analyses, especially by visual inspection of changes in the bone tissue of some joints. These methods were developed by correlating the progressive morphological changes that occur in these articulations with the known age of individuals in these skeletal collections. For example, an analysis of macroscopic changes of the pubic symphysis with age was first published in 1920 by Todd and led to one of the earliest age estimation standards. Since Todd, several age at death estimation methods focusing on the metamorphosis of the pubic symphysis have been described (e.g., Brooks and Suchey 1990; Chen *et al.* 2008; Hanihara and Suzuki 1978; Hartnett 2010; McKern and Stewart 1957; Meindl *et al.* 1985). Other age estimation methods have been established by focusing on other skeletal indicators and using microscopic techniques. However, more recently, the focus of research in the age estimation field has changed towards learning more about the skeletal ageing process by investigating the factors that affect the degenerative changes over time, such as body size and disease status. Understanding the skeletal ageing process and

---

[1] The aim of age estimation methods is to estimate the chronological age an individual had at time of death (calendar age – usually the number of years from birth to death) by observing the biological changes that occur in the skeletal with age (biological age) (Cox 2000). Usually the biological age does not closely correspond to the chronological age (Cunha *et al.* 2009), especially for adults, since skeletal ageing may occur at a faster or slower rate, when compared to actual chronological age.

the factors that affect it in adults is crucial to improve the accuracy and precision of age at death estimation methods (Campanacho 2016; Paine and Brenton 2006). Thus, identified skeletal collections from different countries are a valuable research resource to 1) document age-related changes in the skeleton; 2) establish and test age at death estimation methods; 3) understand the effects of environmental and potential genetic factors that affect skeletal ageing; and 4) compare ageing manifestations and variability across populations. Having additional biographical information about the individuals besides their age at death, such as occupation, cause of death, cadaveric weight and stature, is what makes identified skeletal collections crucial to further our understanding of the skeletal ageing process, something that is not possible to achieve with archaeological collections of unidentified human remains.

The effects of environmental factors on the rate of bone ageing

Age estimation methods when applied to different populations usually yield results that are less accurate than when applied to the original population used to develop the method. This reduction in accuracy may be related to observation error, and to age biases in the collections, but one increasingly used explanation is variation in bone degeneration between populations. It has been suggested that environmental factors, such as occupation, disease, and body size may be responsible for the expression of variation in bone ageing rate between individuals and between populations (Campanacho 2016; Mays 2012; Merritt 2014, 2015; Taylor 2000; Wescott and Drew 2015). However, very little is known about the actual influence of these genetic and environmental factors on skeletal ageing. The lack of a better understanding of the influence of genetic and environmental factors on the ageing process in the skeleton of adults is thought to be the main reason behind the inaccuracy of adult age estimation methods (Jackes 2000). Few studies have dealt with the question of what factors may be affecting the rate at which bone tissue degenerates. A review of the literature shows that the direct or indirect effects of a number of factors have been examined, specifically pregnancy and parturition, drug and alcohol use, occupation and physical activity, disease, diet, and body size or mass. The majority of studies investigating the effect of environmental factors on skeletal ageing have been carried out using identified skeletal collections because of their intrinsic advantage of including individuals of known age at death. In other cases, studies relied on samples of the sternal extremity of the fourth rib and of the pubic symphysis collected from autopsy (Taylor 2000, Hartnett 2007; Passalacqua 2014), rather than relying on traditional dry bone material.

Pregnancy and parturition

It has been suggested that degenerative differences in the pubic symphysis and the auricular surface, two joints in the pelvis, across the sexes may be caused by pregnancy and parturition in women (Igarashi *et al.* 2005). Since a relaxation of the ligaments under the influence of hormones is a natural occurrence during pregnancy, this increases these joints' mobility to facilitate birth (Alicioglu *et al.* 2008; Becker *et al.* 2010; Brooke 1924; Walker 1992). As a consequence, it is possible that pregnancy and childbirth may affect joint degeneration. The effects of parity on the metamorphosis of the pubic symphysis were examined by Hoppa (2000), using the Spitalfields collection, where parity status was available from parochial Baptism Registers (Molleson *et al.* 1993). Hoppa (2000) compared the variation of mean age for state, using the Brooks and Suchey (1990) method, between low-birth vs. high-birth females. No significant difference in mean age for state between the low-birth vs. high-birth female groups was found.

Drugs and alcohol use

The effects of drug or alcohol use on the rate of bone degeneration were examined by Taylor (2000), Hartnett (2007) and Passalacqua (2014). In these studies, it is hypothesized that substance use may affect the metamorphosis of the joints, because it affects homoeostasis and bone mineral density (Hartnett 2007; Passalacqua 2014; Taylor 2000). A decrease in bone mineral density levels have been reported in conjunction with chronic consumption of alcohol and opiates, with alcohol having a toxic effect on osteoblasts (Hartnett 2007; Taylor 2000). Excessive alcohol consumption may also affect hormones' plasma levels accountable for calcium homeostasis (Hartnett 2007; Taylor 2000). Similarly, opiates use may affect the parathyroid hormone secretion (Taylor 2000). Additionally, osteosclerosis, osteomyelitis and septic arthritis — which can affect the skeletal tissue — may result from the intra-venous use of drugs (Hartnett 2007; Taylor 2000). Taylor (2000) observed 173 pairs of the sternal end of the fourth rib extracted from autopsies at the King County Medical Examiner's Office in Seattle, Washington, USA, where the identity of the individuals was known. The morphology of the sternal rib has been shown to change with age (due to degenerative processes) and the types of changes are recorded to score a phase to categorise this metamorphosis (İşcan *et al.* 1984, 1985). Additional information was available as to whether the individual was a chronic drug and alcohol user or not. This was obtained from medical records, from family and friends, and from autopsy findings corroborated by death scene investigations. Two groups of individuals were compared: 65 individuals were chronic substance users while 55 were not. Results suggest chronic substance use may have an effect on the accuracy of the İşcan *et al.* (1984) and İşcan *et al.* (1985) rib age methods, whose morphology also changes with age. Kappa values, a statistical approach used here to test the relationship between degeneration and age, showed marginal agreement between estimated phase and actual phase for both groups (with substance use: k= 0.077; without substance use: k= 0.220). Actual phase indicates the phase the individual supposedly should be placed in according to its real age at the time of death, which may or not coincide with the estimated phase for İşcan *et al.* (1984) and İşcan *et al.* (1985) age estimation method according to the observed metamorphosis of the sternal end of the rib.

In a similar study by Hartnett (2007), the pubic symphysis and sternal ends of the fourth ribs of individuals with known age, sex and ancestry data were collected at the Maricopa County Forensic Science Center in Phoenix, Arizona (USA) and through body donations to the Barrow Neurological Institute, also in Phoenix. Two groups of individuals of both sexes were compared, with (n=99) and without a known history of drug and/or alcohol addiction (n=99). Age at death was estimated by employing the İşcan and Loth (1986a, 1986b) and Brooks and Suchey (1990) methods, as well as the Hartnett-Fulginiti (Hartnett 2007) phase descriptions (a revised version of both previous methods). The mean differences between estimated phase and the actual phase were compared for both groups with a two-sample sign test for related samples. No significant differences in estimated and actual phases between both groups was obtained showing substance use has no influence on age estimation based on the fourth sternal rib ends and pubic symphysis.

Passalacqua (2014) using the Maricopa County Forensic Science Center collection reassessed the results published by Hartnett (2007). Correlations between pubic symphysis age and age at death on one hand, and sternal rib end age and age at death on the other, were calculated for individuals with (n= 94) and without a history of drug/ alcohol use (n= 483). Similar correlation values were obtained between pubic symphysis/ rib end age in both groups of individuals, suggesting that drug and/or alcohol use did not affect the rate of degeneration with age.

Occupation and physical activity

The influence of occupation and physical activity on the rate of pubic symphysis degeneration with age was investigated by Campanacho *et al.* (2012), Mays (2012), and Miranker (2015) to determine whether a more physically demanding occupation and/ or active life-style and consequent higher mechanical stress on the joints may lead to faster bone degeneration at the joints. Campanacho and colleagues examined 161 male individuals from the Lisbon (Cardoso 2006) and Coimbra (Rocha 1995) identified skeletal collections both from Portugal, with known occupation at the time of death. Occupation was coded as manual (n= 93) or non-manual (n= 68) based on the profession stated in the obituary records, and the criteria outlined by Armstrong (1972), Roque (1988), and Alves-Cardoso and Henderson (2010). In the same sample, physical activity was quantified by calculating the femoral robusticity index (Olivier and Demoulin 1984): individuals were coded as robust (performed higher physical demanding activities: n= 64) and gracile individuals (performed lesser physical demanding activities: n= 73) if they were above or below the average robusticity index, respectively. Fourteen age-related morphological traits in the pubic symphysis were scored using a binary scoring system (absence or presence) and differences in the rate of degeneration between the non-manual and manual groups were examined by comparing the transitional ages in the two groups for each trait with a logistic regression model. The analysis was repeated a second time for the physical activity groups (robust *versus* gracile individuals). Only one type of degeneration, a change to the ligament attachments on the ventral bevelling for gracile individuals presented a significantly older mean age of transition between stages suggesting a slower degeneration rate compared to robust individuals. Thus the appearance of bone outgrowths at the ventral bevelling took longer to emerge in gracile individuals compared to robust individuals. The results suggest that occupation and physical activity did not have a significant effect on the degeneration of the pubic symphysis in this sample.

The effects of occupation on bone degeneration were further examined by Mays (2012) in the pelvic part of the hip joint, the acetabulum. The sample used consisted of 50 male individuals from the Spitalfields collection, with known occupation (Molleson *et al.* 1993).

Two groups of individuals were established: non-manual (n= 17) and manual workers (n= 33). A revised scoring system for four acetabular traits significantly correlated with age was adapted from Rissech and colleagues' (2006) age estimation method. The four acetabular trait scores were summed to create a composite score. Manual workers had lower composite scores-for-age compared to non-manual individuals showed by significant differences in the elevation of the regression lines, suggesting slower acetabular degeneration with age for individuals with more physically demanding professions (manual workers).

Miranker (2015) also studied the effects of occupation on age-related degeneration in the pelvic joints (Brooks and Suchey 1990; Calce 2012; Osborne and colleagues 2004; Rissech and colleagues 2006). A total of 203 individuals of both sexes, with known occupation, were studied from the William Bass Donated Skeletal Collection, University of Tennessee, USA (Jantz and Jantz 2008). The individuals were also divided into non-manual and manual groups. The rationale behind this comparison is that a faster degeneration of the pelvic joints will be detected in individuals of manual occupation due to higher biomechanical stress, as a result of their more physical demanding occupations. For the pooled sexes sample, a significant relationship between occupation and the age estimation methods developed by Brooks and Suchey (1990) and Calce (2012) were obtained. Significant results were also obtained but only for female individuals using the method developed by Osborne and colleagues (2004). Age estimation regression lines showed an overestimation of age for non-manual labour compared to manual labour, which is contrary to expectations. As pointed out by Miranker (2015), these results may be questionable, since out of 76 female individuals, only 7 belonged to the manual group.

Disease and diet

Disease and dietary deficiencies may affect bone mineral density, which in turn can also influence skeletal changes associated with ageing. The possible effect of disease and diet on skeletal ageing was investigated by Taylor (2000), Mays (2012) and Paine and Brenton (2006).

Taylor (2000) examined the effects of cardiac and pulmonary diseases on the accuracy of the İşcan and colleagues' (1984 and 1985) methods of age estimation based on rib metamorphosis. Kappa values showed marginal agreement between estimated phase and actual phase for individuals with and without thoracic diseases (with thoracic diseases: n= 63, k= 0.042; without thoracic diseases: n= 88, k= 0.180). Taylor (2000) concluded that thoracic diseases seem to have a noticeable effect on the accuracy of the İşcan and Loth method of age estimation.

Mays (2012) investigated the effects of diffuse idiopathic skeletal hyperostosis (DISH)[2] on the metamorphosis of the acetabulum. The age-related changes in the acetabulum

---

[2] Diffuse idiopathic skeletal hyperostosis, also known as Forestier-Rotés disease, affects the locomotor system through bone proliferation at the axial skeletal (Campillo 2001). A higher frequency of diffuse idiopathic

involve new bone formation at or near the joint margin, which may not be uniquely associated with degeneration related to the ageing process. New bone in the acetabulum may actually reflect a general tendency toward bone formation seen in individuals with DISH. Mays (2012) compared three groups of individuals: those with DISH (three females and nine males), those with subclinical DISH[3] (21 females and 27 males) and those without DISH (60 females and 35 males). Statistical testing showed no differences for the composite acetabular score based on age between the two groups. These results suggest a lack of an effect of DISH and subclinical DISH in the degeneration process of the acetabulum with age, although the DISH group is small.

Metabolic disorders and dietary deficiencies may also cause higher or lower bone turnover levels compared to what it is considered to be normal, thus potentially affecting bone degeneration at the joints. Paine and Brenton (2006) assessed the effect of nonspecific general malnutrition and niacin deficiency (Pellagra) on the reliability of age estimations based on the Stout and Paine (1992) rib method. Paine and Brenton (2006) studied 26 individuals from the Raymond Dart Skeletal Collection and found that, on average, affected individuals were under-aged by 29.2 years (age at death range: 16 to 89 years, mean age at death: 50.2 years; estimated age range: 13.5 to 29.6 years, mean estimated age: 21.0 years). In this study they also found that Haversian canals (a network of longitudinal channels within the structure of bone for the passage of small blood vessels, nerves and lymph vessels) were significantly larger and the cortical area was smaller compared with the control group collected by Stout and Paine (1992) from the Boone County Medical Examiner's Office in Missouri. The large underestimation of age in individuals who died of malnutrition/nutritional deficiency suggests that age estimation standards established from healthy cases may not be applicable to individuals with poor diet and health.

Body size

Biomechanical stress at joints may be higher for taller and heavier individuals due to their larger mass, potentially causing faster degeneration of joints with age. The effects of stature and weight on the rate of degeneration of the pubic symphysis and auricular surface of the ilium were studied by Merritt (2014, 2015). Merritt used the same dataset, which included 764 individuals of both sexes from the Hamann-Todd and William Bass Donated Skeletal Collections, to assess the effects of stature and body mass, but employing different methodological approaches. Merritt (2014) investigated whether cadaveric stature and weight affected the bias and inaccuracy of eight age estimation methods. Ages of shorter and lighter individuals were underestimated, whereas taller and heavier individuals tended to be overestimated (Merritt 2014). In the subsequent 2015 paper, the age-of-transition between phases of eight age estimation methods were established for

---

skeletal hyperostosis occurs in male individuals. Ossification occurs at the periosteum, tendon and ligament insertions, affecting mostly the vertebral column and the pelvic bones. This ligament ossification on the vertebral column has the appearance of dripping candle-wax on the anterior or anterior-lateral face of the vertebral bodies (Campillo 2001).

[3] Mays (2012) classified as subclinical diffuse idiopathic skeletal hyperostosis as an earlier stage of the pathology when individuals exhibited insufficient ossification in the anterior longitudinal ligament.

individuals of different body size values based on cadaveric stature and weight as well as femoral measurements taken from the skeleton. Merritt (2015) found the transition between phases occurred at an earlier age for shorter and lighter individuals. Heavier and taller individuals had significantly higher auricular surface trait scores compared to lighter and shorter individuals. An exception was encountered for macroporosity (1 mm or larger pores leading into the bone structure), with significantly higher scores in individuals whose stature was smaller than 1.61m and for individuals weighing 54.37 kg or less. Surface texture also showed the inverse relationship with lower scores found in individuals with a femoral length less than 430 mm (smaller individuals) and femoral head diameter less than 43 mm (lighter individuals).

Wescott and Drew (2015) also compared the inaccuracy and bias of the Brooks and Suchey (1990) and Buckberry and Chamberlain (2002) age estimation methods between two groups of individuals with normal body mass index (BMI: 18.5–24.9) and obese individuals (BMI ≥ 30). The sample consisted of 226 individuals of both sexes from the William Bass Donated Skeletal Collection. For obese individuals, the age was overestimated with a greater inaccuracy and bias compared to normal body mass individuals. However, differences among BMI groups were only significant for the Buckberry and Chamberlain (2002) method. A higher inaccuracy and bias for the obese individuals was obtained, except for individuals older than 69 years of age for the auricular surface. Additionally, earlier ages-at-transition between estimated phases were found, suggesting an accelerated degeneration rate in obese individuals with age. Similarly, for the pubic symphysis, age estimation's inaccuracy and bias were higher for obese individuals, except for individuals older than 69 years old. An earlier age-at-transition from phase I/II to III occurred for the obese group than for the normal BMI group, suggesting a decelerated ageing rate in normal body mass index individuals for the initial phases, but not for older phases. Similar ages-at-transition were obtained for pubic symphysis older phases between both BMI groups.

Campanacho (2016) investigated if body size—assessed by measurements of stature, body mass, bone robusticity and joint surface area—influenced age-related criteria of the pubic symphysis, auricular surface of the iliac and the acetabulum. Two identified skeletal collections were studied, one from the University of Coimbra (n= 317), along with the William Bass Donated Skeletal Collection from the University of Tennessee (n= 236). Analysis indicated that some of the age-related criteria are affected by body size variables, especially by stature, body mass and joint surface area[4]. This research suggests that individuals with larger skeletal proportions seem to age faster on their pelvic bone joints. However, different patterns of association were found for each collection. Differences between samples may result from bone degeneration rate variability reflecting population differences in lifestyle and environment, reflecting in body size dissimilarities between populations, with the American being taller and heavier than the Portuguese. The only exception is that robusticity has the least effect in ageing for both samples.

---

[4] See Campanacho (2016) thesis for a detailed age-related criteria description and illustrations.

Limitations of identified skeletal collections and their constraints on age estimation research

Identified skeletal collections are crucial for developing age estimation techniques and understanding the relationship between bone degeneration and chronological age. However, identified skeletal collections are not perfect datasets (see also Chapters 4 and 5). For example, degenerative changes observed on the joints of skeletons represent a static reality. It is only possible to observe the modifications present at death and not how or when the changes occurred throughout an individual's life time. Although any other skeletal collections will represent a similarly static reality, there are certain limitations which are intrinsic to these collections and affect how data is sampled and analysed, that require consideration by those undertaking skeletal age-related research. Some of these limitations include age and sex biases, the quality of the biographical records, and the socioeconomic nature of collections themselves.

Due to various factors, individuals in a population do not have equal chances of becoming part of an identified skeletal collection after death. One of the consequences is that those factors will often create biases in the age and sex distribution of the collections. The origin of the skeletal remains (e.g., cemeteries, autopsies and body donations), as well as the legal systems that regulate the collection of human remains, together with health system policies, and associated socio-cultural beliefs concerning the use of the body for research, will determine who will become part of an identified skeletal collection (Albanese 2003; Komar and Grivas 2008; Usher 2002, see also Chapter 4). For example, individuals with infectious diseases, such as hepatitis, HIV, hepatitis and antibiotic resistant infection, will not be accepted onto body donation programs, and thus will not become part of the William Bass Donated Skeletal Collection, except if cremated (Jantz and Jantz 2008; University of Tennessee 2017). Such criteria will limit the representation of infectious diseases in those collections. Thus in future it will not be possible to study the potential skeletal manifestations of certain infectious diseases, including the effects of such conditions on age-related morphological changes in the skeleton. However, investigating the impact of certain infectious diseases, such as tuberculosis, can still be carried out with historic identified collections, but these are often restricted to a pre-antibiotic era. The fact that identified skeletal collections include an amalgamation of cohorts over a long period of time means that is often impossible or difficult to study the effects of certain factors on age-related changes in the skeleton when they vary significantly over long periods of time.

Lack of space and curation costs are other problems faced by museums which house identified skeletal collections. This can impact on the size, diversity and growth of collections, but also limits access to them. The number of individuals that comprise each collection is variable, from 156 individuals curated at the Autonomous University of Yucatan, United Mexican States (Albertos-González *et al.* 2013), to more than 1700 individuals in collection curated at the National Museum of Natural History and Science, Portugal (Cardoso 2006). Larger collections are typically a better resource to understand variation in the skeletal ageing process by allowing sampling of the collection in diverse ways and by controlling different sources of variation. As noted by Albanese (2003) there must be caution in collection sampling, and the demographic data and historical details must be taken into consideration (see also Chapter 4 in this volume by Albanese). A similar sex and age distribution should be established for the purpose of creating and testing of age estimation methods, but this is also key when age-related skeletal processes are being compared between groups within the same population. These intra-collection studies represent an additional challenge because they require very large collections. Some collections are notoriously biased and the difficulty in overcoming this probably explains why they have not been used regularly. For example, the George S. Huntington Anatomical Skeletal Collection at the Department of Anthropology, National Museum of Natural History, Smithsonian Institution, has only 11 subadults, and 2988 adults, in which only 824 are adult females[5]. Despite its size and the theoretical possibility of sampling around the age and sex biases, other problems include the need to independently confirm research results using other larger sample sizes, and the effects of limited and incorrect biographical information.

Some of the studies reviewed in this chapter were performed using small samples, which typically require confirmation from larger samples where more robust results can be obtained. For example, Paine and Brenton (2006) analysed only twenty nine individuals, and it may be beneficial to repeat the study with a larger sample, although histological analyses with identified skeletal remains tends to be very restricted due to its destructive nature. Due to the importance and limited ability to increase the number of identified skeletons in the collections the application of destructive techniques has been restricted and is usually not authorized. Nevertheless, it may be possibly to replicate some of these studies using larger samples from autopsies if ethical approvals can be obtained and such samples are developed. An additional, and more general, limitation of these collections is that the majority were assembled in Europe and North America, and therefore, most research on age-related changes in the skeleton was established on samples from this restricted geographical region. For various reasons this research may not reflect what is happening elsewhere (Schmitt 2004).

Another issue is the quality of the biographical records. The documented age at death data in the Hamann-Todd collection, for example, has been shown to be inaccurate for several individuals, because the age of the cadavers was estimated from soft tissue analysis without supporting documentary evidence (Cox 2000, see also Chapter 5). This imprecision may cause a random error in the age estimation method (Meindl *et al.* 1990, Cox 2000; Katz and Suchey 1989) and it will not provide accurate information about the skeletal ageing process. Additionally, age at death may be self-reported in some of the body donations that were incorporated in certain collections. Therefore, reliability of age at death should be verified with other documents and sources if possible, such as birth, marriage and death certificates and hospital records if available and accessible.

Another limitation is that, despite the often rich biographical information, not all of the individual's life history is known. Because the available information refers only to a brief span of time before and around the time of death, it is impossible to understand

---

[5] Personal communication by Dr David Hunt.

the weight of certain factors on the expression of certain skeletal manifestations of age. For example, cadaveric weight and stature do not account for body size fluctuations that individuals experience over a lifetime (Merritt 2015). Likewise, the occupation stated on records may only report the last known profession of the deceased and may not account for changes in occupation during life or other types of physical activities that individuals may have engaged in during their lives (Campanacho *et al.* 2012; Mays 2012; Vidal 2004; Henderson *et al.* 2013). As for the examination of the potential effects of disease, identified skeletal collections have very limited if any information about medical history besides the cause of death stated in the obituary records. The time period the individual suffered from a certain disease, even if it is the primary cause of death, is almost always unknown. The available biographical data also varies according to the source of the skeletal material that forms the collections, such as cemeteries and body donations. Specimens exhumed from historical and modern cemeteries may have less biographical information derived from the cemeteries records, although there is some potential to gain access to additional information from sources, such as medical records, and birth, baptism and marriage certificates. For example, Santos (2000) collected patient files from the Coimbra University Hospital for 236 individuals in the identified skeletal collection at the University of Coimbra to get a longer term picture of their health. Body donations, on the other hand, have the potential to include more and more detailed biographical information from each individual, but it will be dependent on the diligence of the curators and the will of the donator and his or her family.

Despite the limitations highlighted above, identified collections naturally yield unique biographic information than cannot be found for most archaeological human remains. Documentary evidence, as described above, from throughout an individual's life can provide key insights to further our understanding of the influence of genetic and environmental factors in the skeletal ageing process.

Ethical and legal issues with identified skeletal collections: a Portuguese example

Identified skeletal collections have been of paramount importance to osteological research in Portugal, including the development and testing of age at death estimation methods (Rissech *et al.* 2006; Rougé-Maillart *et al.* 2009), and to further understand ageing in adults (Campanacho *et al.* 2012). Identified skeletal collections in Portugal derive largely from municipal cemeteries and are amassed with the permission from local City Halls. In the cemeteries of Portuguese larger cities, after a legal period of three years, human remains are exhumed from temporary graves and typically placed in secondary interments: reflecting a century-long practice of space management in cemeteries. Since secondary interments can also be temporary and are retained by paying a periodical fee to the cemetery, remains can be considered abandoned if fees are not paid on a regular basis. In these situations the public is informed, by advertisement, that the secondary interments will be cleared if families further delay payment. Once secondary interments are cleared they will remain under the custody of the local municipality and either reburied in communal graves or destroyed in crematoria. This practice has been regulated since the 19th century by Portuguese legislation which establishes that jurisdiction over the fate of unclaimed cemetery remains belongs to the municipalities.

With permission from municipalities, these unclaimed remains have been transferred for long term curation to the Universities of Lisbon, Coimbra, Évora and Porto for research and teaching purposes. Although municipal cemetery regulations (decree- law 411/98, of December 30th) do not specifically mention that unclaimed remains can be used for research and teaching purposes, this is in part regulated by another decree-law (274/99, of July 22nd), which regulates body dissection and extraction of body parts for research. It is worth nothing, however, that the Portuguese cultural heritage legislation does not regulate the amassing of these collections, but it does legislate over the management and use of existing collections.

Recently, Alves-Cardoso (2014) has voiced concerns about a lack of more specific Portuguese legislation regarding the creation of identified skeletal collection from cemetery remains. Although clarification is certainly important, specific legislation seems unwarranted, particularly since general principles and practices can be extracted from the current legislation. Perhaps other concerns should also be highlighted. Access policies and research agreements contribute to respectful and careful handling of the skeleton remains and their individual information, as well as their long term curation and care. Given some of these concerns, a case can also be made with respect to the need for an ethics commission that oversees the curation and use of identified skeletal collections. Within the Portuguese physical anthropological community there is no tradition for studies that involve these collections to be approved by ethics commissions. Of especial concern is the potential sharing of skeletal and/or biographical information in social media, an event witnessed by one of the authors, an aspect not yet considered in current access norms or by curators. Social media is becoming a very fast network for sharing of scientific information, which often includes imagery and untreated data. The sharing of biographical information in social media, or any other hard copy of digital media, poses a serious threat to the duty of protecting personal information which is of the responsibility of the curators. Strict rules about the use of biographic information and avoiding sharing personal identifiers, such as names or addresses, should be a concern shared by curators of these collections. The acquisition and use of photographs and three-dimensional digital replicas of the remains, and their subsequent dissemination in social media should also be discussed by ethical commissions who can advise on appropriate policy.

Conclusion

More research is necessary to understand bone degeneration in adults and its association with age. While considerable knowledge has now accumulated concerning how age-related skeletal features change over an individual's life time, there is very limited understanding of what influences the rate of change with age. Few studies have investigated the effects of various environmental factors, such as occupation and physical activity or disease and health status, on bone degeneration. Therefore, more studies are necessary to further understand the effect of those factors and expand that knowledge to other under-studied populations, particularly outside of Europe and North America. Identified skeletal collections will continue to have a major role on skeletal age-related research due to associated biographical information, particularly that related to age. However, researchers are often unaware of the limitations they face when studying these collections, such as the biases in age and sex distribution and the quality of the biographical information, which can seriously impact and limit research at many levels. Despite these limitations they are still a valuable asset for future age estimation research leading to a better understanding of the skeletal ageing process.

Acknowledgements

We would like to thank the editors for their kind invitation to contribute to this volume and the reviewers for helping to improve the content of this chapter.

References

Albanese, J. 2003. Identified skeletal reference collections and the study of human variation. Unpublished PhD thesis. McMaster University.

Albertos-González, V. M., Ortega-Muñoz, A. and Tiesler, V. G. 2013. A new reference collection of documented human skeletons from Mérida, Yucatan, Mexico. *HOMO – Journal of Comparative Human Biology* 64: 366–376.

Alicioglu, B., Kartal, O., Gurbuz, H. and Sut, N. 2008. Symphysis pubis distance in adults: a retrospective computed tomography study. *Surgical and Radiologic Anatomy* 30: 153–157.

Alves-Cardoso, F. 2008. A Portrait of Gender in Two 19th and 20th Century Portuguese Populations: A Palaeopathological Perspective. Unpublished PhD thesis. Durham University.

Alves-Cardoso, F. 2014. Shaping voids and building bridges: towards an ethical and legal framework and societal approach to Portuguese human identified skeletal collections. Accessed online 08/03/2017

http://cria.org.pt/wp/en/shaping-voids-and-building-bridges-towards-an-ethic-and-legal-framework-and-societal-approach-to-portuguese-human-identified-skeletal-collections-hisc/

Alves-Cardoso, F. and Henderson, C. Y. 2010. Enthesopathy formation in the humerus: data from known age-at-death and known occupation skeletal collections. *American Journal of Physical Anthropology* 141: 550–560

Armstrong, W. A. 1972. The use of information about occupation. In E.A. Wrigley (ed.), *Nineteenth-Century Society: Essays in the Use of Quantitative Methods for the Study of Social Data*: 191–310. Cambridge, Cambridge University Press.

Aykroyd, R. G., Lucy, D., Pollard, A. M. and Roberts, C. A. 1999. Nasty, brutish, but not necessarily short: a reconsideration of the statistical methods used to calculate age at death from adult human skeletal and dental indicators. *American Antiquity* 64: 55–70.

Becker, I., Woodley, S. J. and Stringer, M. D. 2010. The adult human pubic symphysis: a systematic review. *Journal of Anatomy* 217: 475–487.

Boyd, D. C. 1996. Skeletal correlates of human behavior in the Americas. *Journal of Archaeological Method and Theory* 3: 189–251.

Brooke, R. 1924. The sacro-iliac joint. *Journal of Anatomy* 58: 299–305.

Brooks, S. and Suchey, J. M. 1990. Skeletal age determination based on the os pubis: a comparison of the Acsádi-Nemeskéri and Suchey-Brooks methods. *Human Evolution* 5: 227–238.

Buckberry, J. L. and Chamberlain, A. T. 2002. Age estimation from the auricular surface of the ilium: a revised method. *American Journal of Physical Anthropology* 119: 231–239.

Calce, S. E. 2012. A new method to estimate adult age-at-death using the acetabulum. *American Journal of Physical Anthropology* 148: 11–23.

Campanacho, V. 2016. The Influence of Skeletal Size on Bone Degeneration Rate and its Importance in Estimating Age-at-Death in a Portuguese Population (19th-20th centuries) and a North American Population (20th-21st centuries). Unpublished PhD thesis, University of Sheffield.

Campanacho, V., Santos, A. L. and Cardoso, H. F. V. 2012. Assessing the influence of occupational and physical activity on the rate of degenerative change of the pubic symphysis in Portuguese males from the 19th-20th century. *American Journal of Physical Anthropology* 148: 371–378.

Campillo, D. 2001. *Introducción a la Paleopatología*. Barcelona, Edicions Belaterra S.L.

Cardoso, H. F. V. 2006. The collection of identified human skeletons housed at the Bocage Museum (National Museum of Natural History) in Lisbon, Portugal. *American Journal of Physical Anthropology* 129: 173–176.

Chen, X., Zhang, Z. and Tao, L. 2008. Determination of male age at death in Chinese Han population: using quantitative variables statistical analysis from pubic bones. *Forensic Science International* 175: 36–43.

Cox, M. 2000. Ageing adults from the skeleton. In M. Cox and S. Mays (eds), *Human Osteology in Archaeology and Forensic Science*: 61–81. London, Greenwich Medical Media Ltd.

Cunha, E., Baccino, E., Martrille, L., Ramsthaler, F., Prieto, J., Schuliar, Y., Lynnerup, N. and Cattaneo, C. 2009. The problem of aging human remains and living individuals: a review. *Forensic Science International* 193: 1–13.

Cunha, E. and Wasterlain, S. 2007. The Coimbra identified osteological collections. In G. Grupe and J. Peters (eds), *Skeletal Series and Their Socio-Economic Context. Documenta Archaeobiologiae 5*: 23–33. Rahden/Westf, Verlag Marie Leidorf GmbH.

Eliopoulos, C., Lagiab, A. and Manolis, S. 2007. A modern, documented human skeletal collection from Greece. *HOMO – Journal of Comparative Human Biology* 58: 221–228.

Hanihara, K. and Suzuki, T. 1978. Estimation of age from the pubic symphysis by means of multiple regression analysis. *American Journal of Physical Anthropology* 48: 233–240.

Hartnett, K. M. 2007. A Re-Evaluation and Revision of Pubic Symphysis and Fourth Rib Aging Techniques. Unpublished PhD thesis, Arizona State University.

Hartnett, K. M. 2010. Analysis of age-at-death estimation using data from a new, modern autopsy sample – part I: pubic bone. *Journal of Forensic Sciences* 55: 1145– 1151.

Henderson, C. Y., Caffell, A. C., Craps, D. D., Millard, A. R. and Gowland, R. 2013. Occupational mobility in nineteenth century rural England: the interpretation of entheseal changes. *International Journal of Osteoarchaeology* 23: 197–210.

Holliman, S. E. 2011. Sex and gender in bioarchaeology research: theory, method, and interpretation. In S. C. Agarwal and B. A. Glencross (eds), *Social Bioarchaeology*: 149–182. West 217 Sussex, Wiley-Blackwell.

Hoppa, R. D. 2000. Population variation in osteological aging criteria: an example from the pubic symphysis. *American Journal of Physical Anthropology* 111: 185–191.

Hunt, D. R. and Albanese, J. 2005. History and demographic composition of the Robert J. Terry anatomical collection. *American Journal of Physical Anthropology* 127: 406–417.

Igarashi, Y., Uesu, K., Wakebe, T. and Kanazawa, E. 2005. New method for estimation of adult skeletal age at death from the morphology of the auricular surface of the ilium. *American Journal of Physical Anthropology* 128: 324–325.

İşcan, M. Y., Loth, S. R. and Wright, R. K. 1984. Age estimation from the rib by phase analysis; white males. *Journal of Forensic Sciences* 29: 1094–1104.

İşcan, M. Y., Loth, S. R. and Wright, R. K. 1985. Age estimation from the rib by phase analysis: white females. *Journal of Forensic Sciences* 30: 853–863.

İşcan, M. Y. and Loth, S. R. 1986a. Determination of age from the sternal rib in white females: a test of the phase method. *Journal of Forensic Sciences* 31: 990–999.

İşcan, M. Y. and Loth, S. R. 1986b. Determination of age from the sternal rib in white males: a test of the phase method. *Journal of Forensic Sciences* 31: 122–132.

Jackes, M. 2000. Building the bases for paleodemographic analysis: adult age determination. In M. A. Katzenberg and S. R. Saunders (eds), *Biological Anthropology of the Human Skeleton*: 417–466. New York, Wiley-Liss.

Jantz, L. M. and Jantz, R. L. 2008. The anthropology research facility: the outdoor laboratory of the forensic anthropology center, University of Tennessee. In M. W. Warren, H. A. Walsh-Haney and L. E. Freas (eds), *The Forensic Anthropology Laboratory*: 7–22. Boca Raton, CRC Press.

Katz, D. and Suchey, J. M. 1989. Race differences in pubic symphyseal aging patterns in the male. *American Journal of Physical Anthropology* 80: 167–172.

Komar, D. A. and Grivas, C. 2008. Manufactured populations: what do contemporary reference skeletal collections represent? A comparative study using the Maxwell Museum Documented Collection. *American Journal of Physical Anthropology* 137: 224–233.

Lessa, A. 2011. Daily risks: a biocultural approach to acute trauma in pre-colonial coastal populations from Brazil. *International Journal of Osteoarchaeology* 21: 159–172.

Mays, S. 2012. An investigation of age-related changes at the acetabulum in 18th-19th century AD adult skeletons from Christ Church Spitalfields, London. *American Journal of Physical Anthropology* 149: 485–492.

McKern, T. W. and Stewart, T. D. 1957. Skeletal age changes in young American males: analysed from the standpoint of age identification. *Natick: MA Quatermaster Research and Development Command, Technical Report* EP-45.

Meindl, R. S., Lovejoy, C. O., Mensforth, R. P. and Walker, R. A. 1985. A revised method of age determination using the os pubis, with a review and tests of accuracy of other current methods of pubic symphyseal aging. *American Journal of Physical Anthropology* 68: 29–45.

Meindl, R. S., Russel, K. F. and Lovejoy, C. O. 1990. Reliability of age at death in the Hamann-Todd Collection: validity of subselection procedures used in blind tests of the summary age technique. *American Journal of Physical Anthropology* 83: 349–357.

Merritt, C. E. 2014. The Influence of Body Size on Adult Skeletal Age Estimation Methods. Unpublished PhD thesis, University of Toronto.

Merritt, C. E. 2015. The influence of body size on adult skeletal age estimation methods. *American Journal of Physical Anthropology* 156: 35–57.

Miranker, M. 2015. A Test of the Performance of Three Age Indicators of the Adult Human Pelvis and the Influence of Occupation on Morphology. Unpublished MA thesis, New York University.

Molleson, T., Cox, M., Waldron, H. A. and Whittaker, D. K. 1993. *The Spitalfields Project, vol. 2, the anthropology: the middling sort*. CBA Research Report 86. York, Council for British Archaeology.

Olivier, G. and Demoulin, F. 1984. *Pratique Anthropologique à l'Usage des Étudiantes : Ostéologie*. Paris, Université Paris 7.

Osborne, D. L., Simmons, T. L. and Nawrocki, S. P. 2004. Reconsidering the auricular surface as an indicator of age at death. *Journal of Forensic Sciences* 49: 1–7.

Paine, R. R. and Brenton, B. P. 2006. Dietary health does affect histological age assessment: an evaluation of the Stout and Paine (1992) age estimation equation using secondary osteons from the rib. *Journal of Forensic Sciences* 51: 489–492.

Passalacqua, N. 2014. Drug use, homeostasis, and the estimation of age at death from skeletal remains. Poster presented at the *American Academy of Forensic Sciences Meeting*.

Rissech, C., Estabrook, G. F., Cunha, E. and Malgosa, A. 2006. Using the acetabulum to estimate age at death of adult males. *Journal of Forensic Sciences* 51: 213–229.

Rissech, C. and Steadman, D. W. 2011. The demographic, socio-economic and temporal contextualisation of the Universitat Autònoma de Barcelona Collection of identified human skeletons (UAB Collection). *International Journal of Osteoarchaeology* 21: 313–322.

Rocha, M. A. 1995. Les collections ostéologiques humaines identifiées du Musée Anthropologique de l'Université de Coimbra. *Antropologia Portuguesa* 13: 7–38.

Roque, J. L. 1988. *A População da Freguesia da Sé de Coimbra, 1820-1849: Breve Estudo Sócio-Demográfico*. Coimbra, Gabinete de Publicações da Faculdade de Letras.

Rougé-Maillart, C., Vielle, B., Jousset, N., Chappard, D., Telmon, N. and Cunha, E. 2009. Development of a method to estimate skeletal age at death in adults using the acetabulum and the auricular surface on a Portuguese population. *Forensic Science International* 188: 91–95.

Santos, A. L. 2000. A Skeletal Picture of Tuberculosis: Macroscopic, Radiological, Biomolecular and Historical Evidence from the Coimbra Identified Skeletal Collection. Unpublished PhD thesis, University of Coimbra.

Schmitt, A. 2004. Age-at-death assessment using the os pubis and the auricular surface of the ilium: a test on an identified Asian sample. *International Journal of Osteoarchaeology* 14: 1–6.

Stout, S. D. and Paine, R. R. 1992. Histological age estimation using rib and clavicle. *American Journal of Physical Anthropology* 87: 111–115.

Taylor, K. M. 2000. The Effects of Alcohol and Drug Abuse on the Sternal End of the Fourth Rib. Unpublished PhD thesis, Department of Anthropology, University of Arizona.

Todd, T. W. 1920. Age changes in the pubic bone: I. The male white pubis. *American Journal of Physical Anthropology* 3: 285–334.

Umbelino, C., Santos, A. L. and Assis, S. 2012. Beyond the cause of death: other pathological conditions in a female individual from the Coimbra Identified Skeletal Collection (Portugal). *Anthropological Science* 120: 73–79.

University of Tennessee. 2017. Department of Anthropology: College of Arts & Sciences. Available at: http://fac.utk.edu/body-donation/

Usher, B. M. 2002. Reference samples: the first step in linking biology and age in the human skeleton. In R. D. Hoppa and J. W. Vaupel (eds), *Paleodemography: Age Distributions from Skeletal Samples*: 29–47. New York, Cambridge University Press.

Vidal, F. 2004. Factores de diferenciação social em Alcântara no início do século XX: a análise de uma lista de declarações profissionais. *Sociologia, Problemas e Práticas* 45: 53–70.

Walker, J. M. 1992. The sacroiliac joint: a critical review. *Physical Therapy* 72: 903–916.

Wescott, D. J. and Drew, J. L. 2015. Effect of obesity on the reliability of age-at-death indicators of the pelvis. *American Journal of Physical Anthropology* 156: 595–605.

Zuckerman, M. K. and Armelagos, G. J. 2011. The origins of biocultural dimensions in bioarchaeology. In S. C. Agarwal and B. A. Glencross (eds), *Social Bioarchaeology*: 15–43. West 217 Sussex, Wiley-Blackwell.

Chapter 7

Secular changes in cranial size and sexual dimorphism of cranial size: a comparative analysis of standard cranial dimensions in two Portuguese identified skeletal reference collections and implications for sex estimation

Luísa Marinho,[1] Ana R. Vassalo[2] and Hugo F. V. Cardoso[1]

[1] Department of Archaeology and Centre for Forensic Research, Simon Fraser University

[2] Research Centre for Anthropology and Health (CIAS), Department of Life Sciences, University of Coimbra

Introduction

Identified skeletal reference collections have played a crucial role during the history of biological anthropology as a repository of human skeletal and dental variation that has been systematically used for research and teaching purposes. Due to the available documented information about the age, sex, cause or date of death of each individual, together with the often good completeness and preservation of the skeletons, these collections have been used extensively in developing and testing sex and age estimation criteria and methods, for paleopathological identification and interpretation, and as comparative modern references in various human evolutionary studies (Hunt and Albanese 2005). In the forensic literature (Dirkmaat *et al.* 2008; Grivas and Komar 2008; Tomljanovic *et al.* 2006), there has been an increasing concern about the adequacy of these collections for developing and testing sex and age estimation methods, because they typically include individuals who died between the early 19th (or earlier) and early 20th centuries. These collections are thought to be inadequate as a basis for analyzing modern forensic cases and assumed to provide biased estimates of sex, age, or stature due to secular changes in size, health, activity and nutritional status (Dirkmaat *et al.* 2008; Tomljanovic *et al.* 2006). This is a concern that has been expressed in the creation of various new and more modern identified skeletal collections (e.g. Alemán *et al.* 2012; Cardoso *et al.* n/d; Ferreira *et al.* 2014; Sanabria-Medina *et al.* 2016). This same concern can also be applied to archaeological studies because the extent to which these collections represent past populations that differ in size, health and nutritional status, is still largely unknown.

Secular change can be broadly defined as a set of non-genetic cross-generational modifications that occur in the biology and morphology of individuals in a certain population, and that appear in response to changes in living conditions. However, because a significant portion of these changes happen during growth and development, secular trend or change usually refers to an increase in body size and accelerated maturation brought about by an improvement in the material conditions of life which act upon human growth over generations (Eveleth and Tanner 1990). A common explanation for these secular changes is the improved quantity and quality of nutrition, the reduction of the infectious disease burden brought about by immunization and sanitation, as well as increased access to health and medical care (Bogin 1999; Eveleth and Tanner 1990; Malina 1990; Susanne 1984; Tanner 1989). In industrialized nations, this change in the material conditions of life over the last century was sparked by the economic recovery following World War II, with subsequent consequences to human growth (Bogin 1999; Eveleth and Tanner 1990; Malina 1990; Susanne 1984; Tanner 1989). However, not all nations experienced a significant change in living conditions at the same time and in some cases a decline in overall living conditions have resulted in a persistent decrease in body size or delayed maturation (Bogin 1999; Tobias 1985).

The phenomenon of secular change and its implications in the analysis of modern human forensic material is now relatively well known, particularly the effects of increased body size (Case and Ross 2007; Spradley and Jantz 2011) and cranial shape variation (Goode 2015; Jantz 2001; Jantz and Meadows Jantz 2000; Wescott and Jantz 2005) in the estimation of sex, and the effects of increased body size and body shape changes in the estimation of stature (Albanese 2010; Albanese et al. 2012; Kimmerle et al. 2008; Meadows Jantz and Jantz 1995). The effects of improved living conditions have also been examined with respect to age estimation (Langley-Shirley and Jantz 2010), but studies are often restricted to North America. Given the different context for secular change across the world and the distinct origin of identified skeletal collections in various countries, it is important to determine to what extent both older and newer identified collections are representative of the local modern populations, what are the local causes and consequences of the secular change effect, and its implications for the development and testing of osteological methods. This is important because older collections are not necessarily biased and newer ones are not inherently better data sources (See also Albanese in this volume).

This chapter examines the nature and magnitude of the secular change in absolute cranial size over the 20th century in Portugal and its implications for craniometric sex estimation methods. Linear dimensions collected from skulls in two identified skeletal collections originating from the city of Porto that were amassed 100 years apart, are compared in terms of sex and year of birth. An additional purpose is to examine the secular change in the amount of sexual dimorphism of cranial size. Therefore, further to sex-specific absolute changes in size, changes in sexual dimorphism in size are also being evaluated. Changes in sexual dimorphism over time are particularly important for understanding the accuracy of sex estimation methods because they reflect how well males and females can be distinguished. Often overlooked, these variations in sexual dimorphism are also important when examining the effects of secular change because the magnitude of sexual dimorphism is also responsive to changes in the environment. Variation in sexual dimorphism results from females being more buffered against environmental change in terms of their growth and development, due to their reproductive and childbearing role (Brauer 1982). In terms of most anthropometric measures of size, for example, given that females are generally smaller than males, sexual dimorphism will tend to increase in situations of improved living conditions and decrease when they deteriorate. The two hypotheses being examined here are that there were no significant secular changes in absolute cranial size and in sexual dimorphism of cranial size between the two collections, due to their specific demographics and the unique social and economic development of Portugal during the 20th century. Although individuals in the more recent collection were amassed 100 years after the earlier collection, during the 2010s, and despite having died largely within the last 30 years, they were born and grew up before the 1970s, the period that witnessed the greatest improvements in living conditions in Portugal during the 20th century (Cardoso 2008; Padez 2002). Thus, the growth and development of the individuals in the more recent collection antedates the greatest secular change in demographics, health and nutrition experienced by the Portuguese population.

Materials and Methods

Nineteen standard craniometric measurements (Buikstra and Ubelaker 1994) were collected from the cranium and mandible of 130 skulls originating from two Portuguese identified skeletal collections with distinct chronologies. Measurements of the skull are based on the distance between specific anatomical reference points and are used to quantify its size and shape.

The older collection was amassed by Antonio Mendes Correia roughly between 1912 and 1917 and is currently curated at the University of Porto (Natural History Museum and Faculty of Sciences) as the Mendes Correia collection (MCC) (Cardoso and Marinho 2017). Of the 99 individuals of known sex and age comprising this collection, but only 70 skulls were measured for this study due to poor preservation and gaps in the available documentary information. The newer collection was amassed between 2012 and 2014 under the BoneMedLeg research project (BMLC) lead by one of the co-authors (HC) and is currently curated at the Northern Delegation of the National Institute of Legal Medicine and Forensic Sciences, also in Porto (Cardoso et al. n/d). Of the 94 known sex and age individuals, 60 had their skulls measured for this study due to preservation issues. Skeletal remains in both collections originate from unclaimed secondary burial plots at the local cemeteries in the city of Porto (Cardoso and Marinho 2017; Cardoso et al. n/d), the context for this is discussed at the end of this chapter. Although only exact dates of death are known for individuals in both collections, examining secular changes requires comparisons between samples to be done based on dates of birth. Because improvements in living conditions will have their greatest impact during growth and development, and given the diversity in ages at death in the samples, comparisons based on the date of birth are a much better reflection of environmental changes that took place between the time periods of the two samples. Thus, although these collections were amassed 100 years apart, this temporal distance alone might not be enough to capture secular change effects, due to the age of the individuals. Similarly, an analysis based on dates of death is also likely to be based on a time frame that misrepresents those same secular change effects.

Dates of death are not known for the individuals in the MCC, but the remains are known to have been collected within a narrow window of time before 1912 (Cardoso and Marinho 2017). Estimating dates of birth based on known ages at death in the MCC establishes a likely period between 1822 and 1897 (mean year of birth = 1866), or slightly earlier (Cardoso and Marinho 2017). The BMLC was assembled almost exactly 100 years after the MCC, between 2012 and 2014, and includes individuals who died between 1969 and 2001. Dates of birth are not known, but based on known ages at death and dates of death, individuals in this collection were born between 1897 and 1969 (mean year of birth = 1928). A summary of the mean ages at death by sex for both collections is presented in Figure 1. The MCC and the BMLC have their ages at death and years of birth

Collection	Females			Males			Total
	N	Age at death		N	Age at death		
		Mean	SD		Mean	SD	
MCC	36	45.7*	19.97	34	46.2	19.35	70
BMLC	34	63.3**	15.80	26	57.6*	16.93	60
Total	70	53.9	20.04	60	50.9	19.10	130

*Sample includes two individuals of unknown age at death.
**Samples includes four individuals of unknown age at death.

FIGURE 1. NUMBER OF INDIVIDUALS (N) AND MEAN AGE AT DEATH IN YEARS (AND SD) FOR INDIVIDUALS IN THE TOTAL SAMPLE AND SEPARATELY FOR THE MENDES CORREIA (MCC) AND BONEMEDLEG (BMLC) COLLECTIONS.

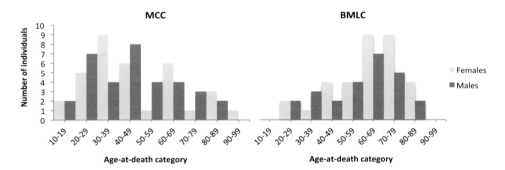

FIGURE 2. AGE AT DEATH (YEARS) DISTRIBUTION FOR THE INDIVIDUALS IN THE MENDES CORREIA (MCC) AND THE BONEMEDLEG (BMLC) COLLECTIONS, BY SEX.

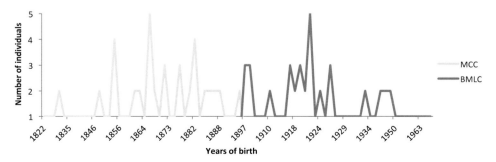

FIGURE 3. YEARS OF BIRTH DISTRIBUTION FOR THE INDIVIDUALS IN THE MENDES CORREIA (MCC) AND THE BONEMEDLEG (BMLC) COLLECTIONS. NOTE THE NON-OVERLAPPING DISTRIBUTIONS AND ALL BIRTH DATES PRIOR TO 1970.

distributions represented in figures 2 and 3, respectively. Place of birth is known for both collections and suggests that both groups of individuals are a fairly good representation of the population of Porto, with many births in similar areas of the city and non-locals from a variety of municipalities from the Northern part of the country and around Porto. None of the collections are known to have targeted specific individuals during the amassing of skeletal remains. However, the BMLC was amassed by selecting those individuals who had a more recent date of death, when compared to the unclaimed set of remains that were available at the cemeteries.

Secular changes in absolute cranial size were assessed by comparing mean values for the 19 craniometric variables between the two collections, using an independent sample t-test for males and females separately. This comparison tested whether male and female cranial measurements have changed or remained the same over the 100-years period. Secular changes in sexual dimorphism of cranial size were evaluated by comparing the amount of sexual dimorphism obtained for the 19 craniometric variables in the two collections. The independent sample t-test is used as the measure of sexual dimorphism, because it quantifies for each variable, not only the distance between the male and female means, but also the amount of overlap between the male and female distributions (Marini *et al.* 1999). The mean of the t-test values was used as a measure of mean sexual dimorphism in each collection. Then, the medians of these t-test values were used as raw data and compared between collections using a non-parametric Mann-Whitney U test to assess changes in sexual dimorphism over time. Sample sizes varied between variables and tests, due to differential preservation of crania and mandibles. All statistical analyses were carried out using the Statistical Package for the Social Sciences (SPSS), version 19.0.

Results

Figures 4 and 5 show the differences in mean absolute cranial and mandibular sizes between the two collections, for females and males, respectively. In the female sample, only basion-bregma height and mastoid process length show statistically significant changes over time, with basion-bregma height increasing and mastoid process length decreasing. In the male sample, only mandibular length shows statistically significant increase over time.

Figures 6 and 7 show the t-test results for the comparison of mean cranial and mandibular measurements between males and females in the MCC and the BMLC, respectively. Twelve out of nineteen measurements show statistically significant differences (p<0.05) between the sexes, but only nine of these measurements were significant in both collections. Both the MCC and the BMLC show statistically significant sex differences in maximum length, maximum breadth, basion-bregma height, base length, nasal height, mastoid process length, mandibular minimum ramus breadth, mandibular maximum ramus breadth and mandibular maximum ramus height. The MCC also shows statistically significant sex differences in the maxillo-alveolar length, upper facial height and mandibular bigonial width, and the BMLC in chin height, mandibular body height and mandibular length. The greatest sexual dimorphism (t-test value>4)

Metric variable	MCC			BMLC			t-test	p
	N	Mean	SD	N	Mean	SD		
Cranial maximum length	27	178.2	5.81	17	176.6	6.38	-0.86	0.398
Cranial maximum breadth	26	131.7	4.58	24	133.8	5.56	1.38	0.175
Basion-bregma height	25	125.6	4.74	19	129.8	7.51	2.31	0.026
Cranial base length	24	95.3	3.56	14	94.7	4.48	-0.47	0.641
Maxillo-alveolar breadth	10	58.9	2.78	15	54.8	13.05	-0.97	0.344
Maxillo-alveolar length	17	50.8	3.22	12	51.9	3.08	0.98	0.337
Upper facial height	18	64.9	3.61	10	67.3	4.58	1.49	0.149
Nasal height	24	48.1	2.44	13	47.9	4.71	-0.14	0.894
Nasal breath	21	22.9	2.07	12	22.5	1.23	-0.78	0.440
Mastoid process length	25	27.6	3.53	23	25.1	5.07	-1.98	0.053
Chin height	12	29.1	2.49	17	29.1	2.97	-0.02	0.988
Mandibular body height	10	27.7	1.73	20	27.5	3.25	-0.23	0.822
Mandibular body breath	18	10.1	1.00	29	10.1	1.30	-0.17	0.865
Mandibular bigonial width	13	92.0	5.42	17	94.6	4.83	1.37	0.182
Mandibular bicondylar breath	13	111.8	3.36	10	112.9	9.12	0.37	0.719
Mandibular minimum ramus breadth	17	29.0	2.42	26	29.1	2.34	0.02	0.981
Mandibular maximum ramus breadth	17	40.8	2.85	21	39.9	3.95	-0.76	0.451
Mandibular maximum ramus height	17	58.8	3.69	21	60.4	3.16	1.46	0.152
Mandibular length	17	77.9	4.17	21	80.2	5.30	1.45	0.156

FIGURE 4. DESCRIPTIVE STATISTICS AND T-TEST RESULTS FOR THE COMPARISON OF MEAN CRANIAL AND MANDIBULAR MEASUREMENTS IN FEMALES BETWEEN THE BONEMEDLEG (BMLC) AND THE MENDES CORREIA (MCC) COLLECTIONS.

Metric variable	MCC			BMLC			t-test	p
	N	Mean	SD	N	Mean	SD		
Cranial maximum length	22	185.7	6.50	18	186.1	7.21	0.20	0.844
Cranial maximum breadth	23	138.2	4.27	21	136.3	3.14	-1.65	0.106
Basion-bregma height	22	132.6	4.15	19	134.1	4.88	1.07	0.292
Cranial base length	22	100.1	3.75	19	101.3	3.94	0.94	0.354
Maxillo-alveolar breadth	17	60.2	3.30	14	58.7	3.31	-1.23	0.229
Maxillo-alveolar length	22	54.1	3.43	9	53.7	1.93	-0.41	0.689
Upper facial height	22	70.3	4.57	9	69.7	5.05	-0.32	0.749
Nasal height	24	51.2	3.69	15	52.3	3.77	0.85	0.399
Nasal breath	24	23.9	2.04	15	23.6	2.28	-0.47	0.639
Mastoid process length	18	30.5	3.30	21	29.2	4.24	-1.10	0.278
Chin height	13	30.9	3.08	14	31.5	2.06	0.57	0.574
Mandibular body height	14	29.2	3.95	16	29.6	2.60	0.33	0.745
Mandibular body breath	18	10.1	1.25	22	10.7	1.26	1.42	0.165
Mandibular bigonial width	10	98.6	5.17	11	98.9	7.66	0.12	0.907
Mandibular bicondylar breath	10	114.8	7.14	10	119.6	5.35	1.70	0.106
Mandibular minimum ramus breadth	18	31.3	3.17	18	31.5	3.89	0.15	0.880
Mandibular maximum ramus breadth	16	43.6	4.18	16	43.6	4.47	0.05	0.963
Mandibular maximum ramus height	16	66.2	6.30	17	66.5	5.53	0.18	0.855
Mandibular length	17	80.6	4.99	18	84.4	5.13	2.23	0.033

FIGURE 5. DESCRIPTIVE STATISTICS AND T-TEST RESULTS FOR THE COMPARISON OF MEAN CRANIAL AND MANDIBULAR MEASUREMENTS IN MALES BETWEEN THE MENDES CORREIA (MCC) AND THE BONEMEDLEG (BMLC) COLLECTIONS.

is found in basion-bregma height, maximum breadth, base length, maximum length, mandibular maximum ramus height and upper facial height for the MCC, and in base length, mandibular maximum ramus height and maximum length for the BMLC. Base length, mandibular maximum ramus height and maximum length are the variables with the highest sexual dimorphism common to both collections. Although basion-bregma height and maximum breadth are the two most sexual dimorphic variables in the MCC, they are the two least sexual dimorphic variables in the BMLC.

As a measure of overall sexual dimorphism the mean t-test value for the MCC is 2.77, but when only the statistically significant variables are considered the mean increases to 3.69. The same mean for the BMLC is 2.40 but increases to 2.91 when only the statistically significant variables are used to calculate it. When all variables are considered, the Mann–Whitney U test does not find any statistically significant differences in median sexual dimorphism between the collections (p=0.440). However, when only the most significant sexually dimorphic variables are compared, the MCC seems to show greater median sexual dimorphism than the BMLC, although the p-value is borderline at the 0.05 level (p=0.059).

Discussion

Results in this study suggest that there was no secular change in absolute cranial size and only a slight reduction in, or unchanged, sexual dimorphism of cranial size. Given that BMLC individuals' birth date and growth period antedate the major social and economic improvements in Portugal that sparked significant changes in the health and nutrition of the population (Cardoso 2008), it is perhaps no surprise that individuals in the BMLC do not show modified cranial dimensions or greater sexual dimorphism. This meets the expectations outlined earlier in this chapter. An alternate explanation, however, for the absent secular change in cranial size is the broad range in years of birth in both collections. Although the mean years of birth in the collections are about 60 years apart, the distribution of years of birth is quite broad and covers several decades in both collections. This may have had a dampening effect on cranial changes over time. The fact that mean age at death in the BMLC is slightly older than that of the MCC may have also dampened the effect further due to age-related changes in cranial size, particularly those related to tooth loss and masculinization of female crania (Walker 1995).

On the other hand, if there was a reduction in sexual dimorphism, results do not show a corresponding greater decrease in absolute size in males, or greater increase in absolute size in females due to masculinization. This suggests that the slight reduction in sexual dimorphism observed might be an artefact of the analysis and that both absolute cranial size in males and females, on one hand, and sexual dimorphism of cranial size, on the other, might actually not be showing any differences between the collections. This again confirms expectations. However, these results do not exclude the possibility of secular modifications in cranial size over the last century, but rather that the individuals in the BMLC are probably too old and were born too early to capture the most significant social, economic, health and nutritional transitions during

Metric variable	Females			Males			t-test	p
	N	Mean	SD	N	Mean	SD		
Cranial maximum length	27	178.2	5.81	22	185.7	6.50	4.26	0.000
Cranial maximum breadth	26	131.7	4.58	23	138.2	4.27	5.14	0.000
Basion-bregma height	25	125.6	4.74	22	132.6	4.15	5.34	0.000
Cranial base length	24	95.3	3.56	22	100.1	3.75	4.46	0.000
Maxillo-alveolar breadth	10	58.9	2.78	17	60.2	3.30	1.05	0.302
Maxillo-alveolar length	17	50.8	3.22	22	54.1	3.43	3.09	0.004
Upper facial height	18	64.9	3.61	22	70.3	4.57	4.06	0.000
Nasal height	24	48.1	2.44	24	51.2	3.69	3.44	0.001
Nasal breath	21	22.9	2.07	24	23.9	2.04	1.61	0.114
Mastoid process length	25	27.6	3.52	18	30.5	3.30	2.76	0.009
Chin height	12	29.1	2.49	13	30.9	3.08	1.67	0.109
Mandibular body height	10	27.7	1.73	14	29.2	3.95	1.22	0.237
Mandibular body breath	18	10.1	1.00	18	10.1	1.25	-0.11	0.915
Mandibular bigonial width	13	92.0	5.42	10	98.6	5.17	2.94	0.008
Mandibular bicondylar breath	13	111.8	3.36	10	114.8	7.14	1.22	0.245
Mandibular minimum ramus breadth	17	29.0	2.42	18	31.3	3.17	2.39	0.023
Mandibular maximum ramus breadth	17	40.8	2.85	16	43.6	4.18	2.25	0.032
Mandibular maximum ramus height	17	58.8	3.69	16	66.2	6.30	4.12	0.000
Mandibular length	17	77.9	4.17	17	80.6	4.99	1.67	0.105

FIGURE 6. *DESCRIPTIVE STATISTICS AND T-TEST RESULTS FOR THE COMPARISON OF MEAN CRANIAL AND MANDIBULAR MEASUREMENTS BETWEEN MALES (M) AND FEMALES (F) IN THE MENDES CORREIA COLLECTION (MCC).*

Metric variable	Females			Males			t-test (M-F)	p
	N	Mean	SD	N	Mean	SD		
Cranial maximum length	17	176.6	6.38	18	186.1	7.21	4.13	0.000
Cranial maximum breadth	24	133.7	5.56	21	136.3	3.14	2.01	0.052
Basion-bregma height	19	129.8	7.51	19	134.1	4.88	2.05	0.048
Cranial base length	14	94.7	4.48	19	101.3	3.942	4.45	0.000
Maxillo-alveolar breadth	15	54.9	13.05	14	58.7	3.31	1.09	0.285
Maxillo-alveolar length	12	51.9	3.08	9	53.7	1.93	1.51	0.149
Upper facial height	10	67.3	4.58	9	69.7	5.05	1.11	0.283
Nasal height	13	47.9	4.71	15	52.3	3.77	2.70	0.012
Nasal breath	12	22.5	1.23	15	23.6	2.28	1.61	0.121
Mastoid process length	23	25.1	5.07	21	29.2	4.24	2.86	0.006
Chin height	17	29.1	2.97	14	31.5	2.06	2.71	0.011
Mandibular body height	20	27.5	3.25	16	29.6	2.60	2.11	0.042
Mandibular body breadth	29	10.1	1.30	22	10.7	1.26	1.61	0.113
Mandibular bigonial width	17	94.6	4.83	11	98.9	7.67	1.85	0.076
Mandibular bicondylar breath	10	112.9	9.12	10	119.6	5.35	1.99	0.061
Mandibular minimum ramus breadth	26	29.1	2.34	18	31.5	3.89	2.40	0.024
Mandibular maximum ramus breadth	21	39.9	3.95	16	43.6	4.47	2.68	0.011
Mandibular maximum ramus height	21	60.4	3.16	17	66.5	5.53	4.28	0.000
Mandibular length	21	80.2	5.30	18	84.4	5.13	2.48	0.018

FIGURE 7. DESCRIPTIVE STATISTICS AND T-TEST RESULTS FOR THE COMPARISON OF MEAN CRANIAL AND MANDIBULAR MEASUREMENTS BETWEEN MALES (M) AND FEMALES (F) IN THE BONEMEDLEG COLLECTION (BMLC).

their growth and development. Also, if the reduction in sexual dimorphism was to be demonstrated or confirmed by other studies, then it would probably not be entirely surprising given the negative secular trend in height documented in Portugal between the late 19th and the early 20th centuries (Reis 2002; 2009; Stolz *et al.* 2013). In this respect, it is interesting to note that Portuguese height only seems to exceed its 1860-80 level (the period when most of the individuals in the MCC were born) after 1920-30 (when most of the individuals in the BMLC collection had already been born) (Stolz *et al.* 2013). This negative secular trend reflects long-term multigenerational poverty effects (See also Albanese in this volume) to which the Portuguese population has only recently emerged from. A final explanation is the possibility that the comparisons made here are unable to detect any existing small differences due to relatively small sample sizes.

Several studies have documented secular changes in cranial size or shape (Buretić-Tomljanović *et al.* 2007; Jonke *et al.* 2007; Smith *et al.* 1986; Susanne *et al.* 1988), but findings described here seem to mirror only those reported by Cameron *et al.* (1990) who also detected little if any changes in cranial size and shape of South African males over a period where no improvements in living conditions were taking place. No changes in absolute size suggest that there are no changes in shape as well, but a more detailed analysis would be needed to confirm such assumption. Results presented here, however, seem to contradict the results of Weisensee and Jantz (2011), who have reported a significant secular change in cranial shape of the Portuguese between 1806 and 1954 using specimens from the Lisbon collection, essentially the same time span covered by this study. Although the differences in results between the present study and that of Weisensee and Jantz (2011) may reflect a different analytical approach in the study of secular changes in cranial shape, there is a significant number of issues with the use and interpretations of data by Weisensee and Jantz (2011) that need to be highlighted. Given that the sample used in Weisensee and Jantz's (2011) study also antedates the period with the most significant positive environmental changes experienced by the Portuguese in the 20th century, what could explain the discrepancy between the two studies? One possibility is the dampening effect of the MCC and BMLC years of birth distribution and the slightly older age of individuals in the BMLC mentioned earlier. Another possibility, however, rests in Weisensee and Jantz's (2011) study, namely in their biased sample selection for the periods before and after 1900, which resulted in a mean age at death for each period sample of 67.8 years and 36.0 years, respectively. Not only is this difference indicative of significant age effects when comparing the earlier and the more recent samples, but it is particularly at odds with an anticipated increase in life expectancy if significant changes in living conditions were taking place, as Weisensee and Jantz (2011) assume. The increase in life expectancy would be reflected on the demographic profile of the collection, because those who died later in the 20th century would be on average older than those who died earlier. Some of the effects associated with comparing samples with a very different age profile are potentially related to masculinization of the female crania, for example, but also to tooth loss and the consequent remodeling of the maxillary and mandibular alveoli, and periodontitis, which are very common in the older individuals of the Lisbon collection. Since the collection will reflect the mortality profile of the population from which it is drawn, this very significant age difference is indicative of selective sampling of the collection, which was done to maximize the range of years of birth, but that may have also had unpredicted consequences. In addition, Weisensee and Jantz's (2011) study might also be overstating results and thus, if that is taken to account, their findings may actually be more aligned with the current study. For example, in the model that Weisensee and Jantz (2011) used to detect secular change effects, year of birth explains less than 1% of the total cranial shape variation. Furthermore, because of the nature of their methodology and poor reporting of results, it is not clear which dimensions changed over time, by how much and in which direction. Finally, some of Weisensee and Jantz's (2011) overstated results seem to fit nicely with their expectations about secular change in Portugal, but the reality is that these expectations are based almost entirely on a broad notion of secular and demographic change in Europe, that does not reflect the idiosyncrasies of Portuguese history. Weisensee and Jantz (2011) assume that the demographic and epidemiological transitions, as well as other major environmental changes behind a secular trend in Portugal, were similar and concurrent to that of other industrialized nations of Western Europe, such as the UK, and the US. This is a misconception about the history of demography and epidemiology in Europe as a whole, and of Portugal in particular. Portugal was one of the poorest Western European countries, and has a long history of catching up to changes occurring in Western Europe at a much later date and pace than other countries (see for example, Cardoso 2008).

Findings reported in this chapter raise some concerns about selecting the most appropriate reference sample when estimating the sex of individuals in a population that underwent (or is undergoing) major changes in living conditions in its recent history. It also highlights some of the potential practical problems that forensic anthropologists can face when dealing with a population that experienced recent transition. For example, when forensic anthropologists are investigating the death of relatively young adults (<40-50 years), these individuals may have been born after or about the same time as a certain threshold period of change or transition in living conditions for the population (the 1960s and 1970s in Portugal, for example). In these circumstances, their cranial morphology may reflect a significant secular change effect as a result of their exposure to the new environments during growth. The sex of these individuals might not be adequately estimated when using methods that are based on earlier reference collections that include individuals whose date of birth and growth period antedate that threshold period of change. Conversely, when sexing older individuals in a forensic investigation (>50 years), earlier samples may actually be more adequate because they are a better approximation of individuals whose dates of birth antedate the threshold period, and have not experienced change during growth. Consequently, newer collections are not without biases and they do not necessarily represent ideal forensic datasets, just because they have been amassed very recently. Other recent samples (Ferreira *et al.* 2014; Gonçalves 2014; Sanabria-Medina *et al.* 2016) have similar demographic profiles, potentially combining different cohorts from the same population undergoing transition or change, and should be carefully considered as 'forensic'. Because these collections tend to include a disproportional number of older individuals, these will be representative of earlier periods that may antedate significant secular changes or transitions. Unlike other parts of the world, there are now several 'older' and 'newer' collections housed in various Portuguese institutions that allow researchers to ask specific questions about the effects of secular change on size, shape, morphology and structure of the skeleton, and its forensic implications. For example, in a study by Gonçalves (2014) it is assumed that the modern sample will inform about secular changes in osteometric dimensions when compared to an earlier sample, without carefully considering the demographics of the 'modern' sample, particularly the years of birth, and the history of political, economic and social change, and health/nutritional transition in Portugal. With an average for years of birth between 1915 and 1930 (calculated from reported years of death and mean ages at death) in Gonçalves' (2014) study, it is unlikely that such a 'modern' sample will reflect any significant secular changes in living conditions (it may reflect other causes) experienced by the Portuguese population, which only occurred after the 1960s and 1970s. Navega and colleagues (2013) have also used a similar sample with overlapping years of birth, which include individuals born before 1970. Although Navega *et al.*'s (2013) study examined the effects of physical activity on the geometric proprieties of adult femora, developmental causes cannot be discarded as a factor and thus the lack of a secular trend effect in their study may also reflect a sample that antedates the major changes in living conditions and health transition which occurred in Portugal after the 1960s. Neither sample should probably be considered a fully-fledged contemporary or forensic collection.

Cemetery samples, such as those that comprise many skeletal reference collections, include individuals who might not be considered 'modern' and are a cross-sectional sample of the general population drawn from a wide range of decades and cohorts. If the population experienced a period of transition, different cohorts might be affected differently by secular change. A more 'modern' sample may include a large segment of older individuals who, despite having died recently, do not necessarily represent the current population as a whole. Rather, they may be more representative of individuals living at an earlier moment of the secular change, or even of individuals who actually antedate that period of change. Consequently, representativeness of samples is a matter that should not be taken lightly and researchers should consider critically the nature of both the forensic (target) and cemetery (reference) samples. Depending on the context of the collections and the purpose of the research, it is likely that both 'modern' and 'older' collections can be adequately sampled to address questions of both forensic and archaeological relevance.

A final consideration is one that concerns ethical and legal issues about the amassing and study of identified skeletal reference collections. The collections used for this study are based on unclaimed remains from the cemeteries in the city of Porto. In these cemeteries, a large number of graves are temporary, where the remains are exhumed after 5 years, if skeletonization is complete. These exhumed remains are then normally placed in secondary above ground internments for a periodical fee. Remains are often deemed abandoned or unclaimed when after several years of decades the families stop paying these fees. According to Portuguese legislation since the 19th century, jurisdiction over the fate of unclaimed cemetery remains falls under the municipalities. Due to space management requirements, these unclaimed remains are cleared from their internments and ultimately incinerated by the cemeteries. However, for more than

100 years, the municipal cemeteries have been providing universities and museums with human skeletal remains for teaching and research purposes in Portugal, under this municipal legislation. Although the municipal cemetery regulations (decree-law 411/98, of December 30th) do not specifically identify or foresee the use of unclaimed remains for teaching and research, another decree-law (274/99, of July 22nd), which regulates body dissection and extraction of body parts, tissues or organs for teaching and research purposes, has provided the legal framework for such use since 1999. On the other hand, because these remains are sourced from managed cemeteries, the amassing of unclaimed remains for identified skeletal reference collections does not fall under the jurisdiction of Portuguese cultural heritage legislation, which covers those from archaeological contexts. This legislation, however, does regulate, to some extent, the management and use of existing collections. The purportedly unclear intersection of these three pieces of legislation (cemetery regulations, use of cadavers and body parts for research, and cultural heritage) has raised concerns about the legality and ethics of amassing cemetery based skeletal reference collections and motivated a plea for the creation of specific legislation (Alves-Cardoso 2014). This appeal seems excessive, particularly since the legal framework for these collections can be drawn from general principles, provisions and practice of the existing legislation by analogy, as is common in various areas of the law. Interestingly, the National Ethics Commission for the Life Sciences in Portugal has recently addressed the legal and ethical concerns with the amassing of unclaimed remains for research collections, in a public document published on-line (CNECV 2015). This document was generated for very specific purposes, but because its recommendations are drawn from general existing legal principles by analogy it has effectively created a general legal and ethical framework for the amassing and conservation of these collections in Portugal, thus deeming unnecessary and even redundant any formal changes in legislation. Although relevant legislation exists, it has just never been framed properly.

The principles of dignity, respect, privacy, right to a decent disposal of the remains by the next of kin and not to violate the last known desire of the deceased are paramount (Holland 2015; Walker 2008). However, because remains incorporated in collections are unclaimed by relatives, several of these rights are eventually transmitted or forfeited to municipalities, which have to dispose of the remains in some way. In these circumstances, a more general principle which recognizes the descent community authority to control the disposition of the remains should be followed. Finally, owing to their scientific importance, the long-term preservation and availability of such collections for research purposes is also an ethical imperative, one which is illustrated by the contributions in this book. In practice, however, the challenge arises from trying to reconcile all of these principles. Additionally, perhaps as important as establishing a clear legal and ethical framework for amassing these collections, is instituting a system of ethics reviews and approvals that oversee the use and conservation of existing collections and which have never been traditionally in place among the Portuguese physical anthropology community.

References

Albanese, J. 2010. A critical review of the methodology for the study of secular change using skeletal data. In *Ontario Archaeology No. 85-88/London Chapter OAS Occasional Publication No. 9*: 139–155.

Albanese, J., Osley, S. E. and Tuck, A. 2012. Do century-specific equations provide better estimates of stature? A test of the 19–20th century boundary for the stature estimation feature in Fordisc 3.0. *Forensic Science International* 219: 286.e1–286.e3.

Alemán, I., Irurita, J., Valencia, A. R., Martínez, A., López-Lázaro, S., Viciano, J. and Botella, M. C. 2012. Brief communication: the Granada osteological collection of identified infants and young children. *American Journal of Physical Anthropology* 149(4): 606–610.

Alves-Cardoso, F. 2014. Shaping voids and building bridges: towards an ethical and legal framework and societal approach to Portuguese human identified skeletal collections. Research project abstract. Available at:: http://cria.org.pt/wp/en/shaping-voids-and-building-bridges-towards-an-ethic-and-legal-framework-and-societal-approach-to-portuguese-human-identified-skeletal-collections-hisc/

Bogin, B. 1999. *Patterns of Human Growth*. Cambridge, Cambridge University Press.

Brauer, G. W. 1982. Size sexual dimorphism and secular trend: indicators of subclinical malnutrition? In R. L. Hall (ed.), *Sexual Dimorphism in Homo sapiens: A Question of Size*: 245–259. Praeger, New York.

Buretić-Tomljanović, A., Giacometti, J., Ostojić, S. and Kapović, M. 2007. Sex-specific differences of craniofacial traits in Croatia: the impact of environment in a small geographic area. *Annals of Human Biology* 34(3): 296–314.

Cameron, N., Tobias, P. V., Fraser, W. J. and Nagdee, M. 1990. Search for secular trends in calvarial diameters, cranial base height, indices, and capacity in South African negro crania. *American Journal of Human Biology* 2: 53–61.

Cardoso, H. F. V. 2008. Secular changes in body height and weight of Portuguese boys over one century. *American Journal of Human Biology* 20: 270–277.

Cardoso, H. F. V. and Marinho, L. 2017. Lost and then found: the Mendes Correia collection of identified human skeletons curated at the University of Porto, Portugal. *Antropologia Portuguesa* 32/33: 29-46.

Cardoso, H. F. V., *et al.* (n/d) The human skeletal reference collection amassed under the BoneMedLeg research project (Porto, Portugal). (in preparation)

Case, D. T. and Ross, A. H. 2007. Sex determination from hand and foot bone lengths. *Journal of Forensic Sciences* 2(2): 264–70.

CNECV. 2015. Pedido de doação de ossadas e trasladação destas para o Canadá. 85/CNECV/2015. Conselho National de Ética para a Ciências da Vida. Lisboa. Avalable at: http://www.cnecv.pt/admin/files/data/docs/1446826133_Parecer%2085CNECV2015.pdf

Dirkmaat, D. C., Cabo, L. L., Ousley, S. D. and Symes, S. A. 2008. New perspectives in forensic anthropology. *American Journal of Physical Anthropology* 137(S47): 33–52.

Eveleth, P. B. and Tanner, J. M. 1990. *Worldwide Variation in Human Growth. Second Edition.* Cambridge, Cambridge University Press.

Ferreira, M. T., Vicente, R., Navega, D., Gonçalves, D., Curate, F. and Cunha, E. 2014. A new forensic collection of 21st century identified human skeletons housed at the University of Coimbra, Portugal. *Forensic Science International* 245: 202.e1–202.e5.

Gonçalves, D. 2014. Evaluation of the effect of secular changes in the reliability of osteometric methods for the sex estimation of Portuguese individuals. *Cadernos do GEEvH* 3(1): 53–65.

Goode, K. 2015. Secular trends in cranial morphological traits: a socioeconomic perspective of change and sexual dimorphism in North Americans 1849–1960. *Annals of Human Biology* 42(3): 255–261.

Grivas, C. R. and Komar, D. A. 2008. Kumho, Daubert, and the nature of scientific inquiry: implications for forensic anthropology. *Journal of Forensic Sciences* 53: 771–776.

Haas, J., Buikstra, J. E., Ubelaker, D. H. and Aftandilian, D. (eds) 1994. *Standards for Data Collection from Human Skeletal Remains: Proceedings of a Seminar at the Field Museum of Natural History*, 44. Fayetteville, Arkansas Archaeological Survey Research Series.

Holland, T. D. 2015. 'Since I must please those below': human skeletal remains research and the law. *American Journal of Law & Medicine* 41: 617–55.

Hunt, D. R. and Albanese, J. 2005. History and demographic composition of the Robert J. Terry anatomical collection. *American Journal of Physical Anthropology* 127: 406–417.

Jantz, R. L. 2001. Cranial change in Americans: 1850–1975. *Journal of Forensic Sciences* 46: 784–787.

Jantz, R. L. and Meadows Jantz, L. 2000. Secular change in craniofacial morphology. *American Journal of Human Biology* 12: 327–338.

Jonke, E., Prossinger, H., Bookstein, F. L., Schaefer, K., Bernhard, M. and Freudenthaler, J. W. 2007. Secular trends in the facial skull from the 19th century to the present, analyzed with geometric morphometrics. *American Journal of Orthodontics and Dentofacial Orthopedics* 132(1): 63–70.

Kimmerle, E. H., Jantz, R. L., Konigsberg, L. W. and Baraybar, J. P. 2008. Skeletal estimation and identification in American and East European populations. *Journal of Forensic Sciences* 53(3): 524–532.

Langley-Shirley, N. and Jantz, R. L. 2010. A Bayesian approach to age estimation in modern Americans from the clavicle. *Journal of Forensic Sciences* 55(3): 571–83.

Malina, R. M. 1990. Research on secular trends in auxology. *Anthropologischer Anzeiger* 48: 209–222.

Marini, E., Racugno, W. and Borgognini Tarli, S. M. 1999. Univariate estimates of sexual dimorphism: the effects of intrasexual variability. *American Journal of Physical Anthropology* 109: 501–508.

Meadows Jantz, L. and Jantz, R. L. 1995. Allometric secular change in the long bones from the 1800s to the present. *Journal of Forensic Sciences* 40(5): 762–767.

Navega, D., Cunha, E., Lima, J. P. and Curate, F. 2013. The external phenotype of the proximal femur in Portugal during the 20th century. *Cadernos do GEEvH* 2(1): 40–44.

Padez, C. 2002. Stature and stature distribution in Portuguese male adults 1904-1998: the role of environmental factors. *American Journal of Human Biology* 14: 39–49.

Reis, J. 2002. Crescimento económico e estatura humana. Há um paradoxo antropométrico em Portugal no século XIX? *Memórias da Academia das Ciências de Lisboa, Classe de Letras* 35: 153–169.

Reis, J. 2009. Urban premium or urban penalty? The case of Lisbon, 1840-1912. *História Agrária* 47: 69–94.

Sanabria-Medina, C., González-Colmenares, G., Restrepo, H. O. and Rodríguez, J. M. G. 2016. A contemporary Colombian skeletal reference collection: a resource for the development of population specific standards. *Forensic Science International* 266: 577. e1–577.e4.

Smith, B. H., Garn, S. M. and Hunter, W. S. 1986. Secular trends in face size. *The Angle Orthodontist* 56(3): 196–204.

Spradley, M. K. and Jantz, R. L. 2011. Sex estimation in forensic anthropology: skull versus postcranial elements. *Journal of Forensic Science* 56(2): 289–96.

Stolz, Y., Baten, J. and Reis, J. 2013. Portuguese living standards, 1720-1980, in European comparison: heights, income, and human capital. *Economic History Review* 66(2): 545–578.

Susanne, C. 1984. Living conditions and secular trend. *Studies in Human Ecology* 6: 93–99.

Susanne, C., Vercauteren, M., Krasnicanova, H., Jaeger, V., Hauspie, R. and Bruzec, J. 1988. Secular evolution of cephalic dimensions. *Bulletins et Mémoires de la Société d'anthropologie de Paris* 5(3): 151–161.

Tanner, J. M. 1989. *Fetus Into Man: Physical Growth from Conception to Maturity.* Cambridge, Harvard University Press.

Tobias, P. V. 1985. The negative secular trend. *Journal of Human Evolution* 14: 347–356.

Tomljanovic, A. B., Ostojic, S. and Kapovic, M. 2006. Secular change of craniofacial measures in Croatian younger adults. *American Journal of Human Biology* 18(5): 668–675.

Walker, P. L. 1995. Problems of preservation and sexism in sexing: some lessons from historical collections for paleodemographers. In S. R. Saunders and D. A. Herrings (eds), *Grave Reflections: Portraying the Past Through Cemetery Studies*: 31–48. Toronto, Canadian Scholars Press.

Walker, P. L. 2008. Bioarchaeological ethics: a historical perspective on the value of human remains. In A. M. Katzenberg and S. R. Saunders (eds), *Biological Anthropology of the Human Skeleton*: 3–39. Chichester, John Wiley & Sons.

Weisensee, K. E. and Jantz, R. L. 2011. Secular changes in craniofacial morphology of the Portuguese using geometric morphometrics. *American Journal of Physical Anthropology* 145(4): 548–59.

Wescott, D. J. and Jantz, R. L. 2005. Assessing craniofacial secular change in American blacks and whites using geometric morphometry. In D. E. Slice (ed.), *Modern Morphometrics in Physical Anthropology*: 231 –245. New York, Kluwer Academic/ Plenum Publishers.

Chapter 8

Lives Not Written in Bones: Discussing Biographical Data Associated With Identified Skeletal Collections

Francisca Alves Cardoso[1]

[1] LABOH – Laboratory of Biological Anthropology and Human Osteology, CRIA – Centro em Rede de Investigação em Antropologia, Faculdade de Ciências Sociais e Humanas, Universidade NOVA de Lisboa, Portugal.

Introduction

This chapter aims to explore the limitation of using biographical data associated with individuals within identified skeletal collections. It will focus on the information related with occupation as listed at death, proposing an alternative way to deepen the concept of occupation, and occupational related activities. By exploring the constraints associated with biographical data as listed at death (which is representative of a moment in time) this chapter highlights the importance of an ethnographical approach to understand the complexity of the categorization of occupation, and the impact this may have in the categorization of occupation in association with identified skeletal collections. The emphasis will be placed on the occupation listed as *doméstica* (housekeeper) since it is dominant in all Portuguese identified skeletal collections, it is used as an example of all identified collections, and because categorising this occupation represents one of the major challenges of interpretation. The chapter will discuss identified skeletal collections and the limitations of biographical data and the impact this has on how activity is inferred via the analysis of bones. This chapter also aims to contribute to the description of 'What does it mean to be a *doméstica* (housekeeper)', accentuating the assumption that the term *doméstica* (housekeeper) is complex and varied in the number of activities and tasks undertaken. On a side note, the author feels the necessity to state that this research is in a preliminary stage, and it should be regarded as a pilot-study for a larger research project intending to incorporate multiple methodological approaches to the study of activity, activity-patterns and occupation.

Identified Skeletal Collections and Biographical Data

Osteological collections composed of skeletonized remains of individuals with known biographical data are seen by anthropologists, and other scholars dedicated to the study of osteological human remains, as ideal grounds to test and develop methods to reconstruct past health and behaviour. These collections tend to be referred to as identified skeletal collections, or reference collections. In contrast to archaeological collections, in which human skeletonized remains are excavated from archaeological excavations (however see Chapter 2 which discusses archaeologically-derived identified skeletons), identified skeletal collections are composed of complete to almost complete skeletons exhumed from modern cemeteries, remains of autopsied and donated bodies and anatomical collections. Note that some archaeological collections may also be composed of or contain identified skeletal remains, such as the Spitalfields and St Brides Collections (Molleson and Cox 1993; Cox 1996; Scheuer and Black 1995; Scheuer and Bowman 1995).

Biographical data typically includes name of the individual, place of birth, year of birth, place and location of death, sex, occupation, cause of death and address at time of death, among other information as listed at time of death (examples of which can be found

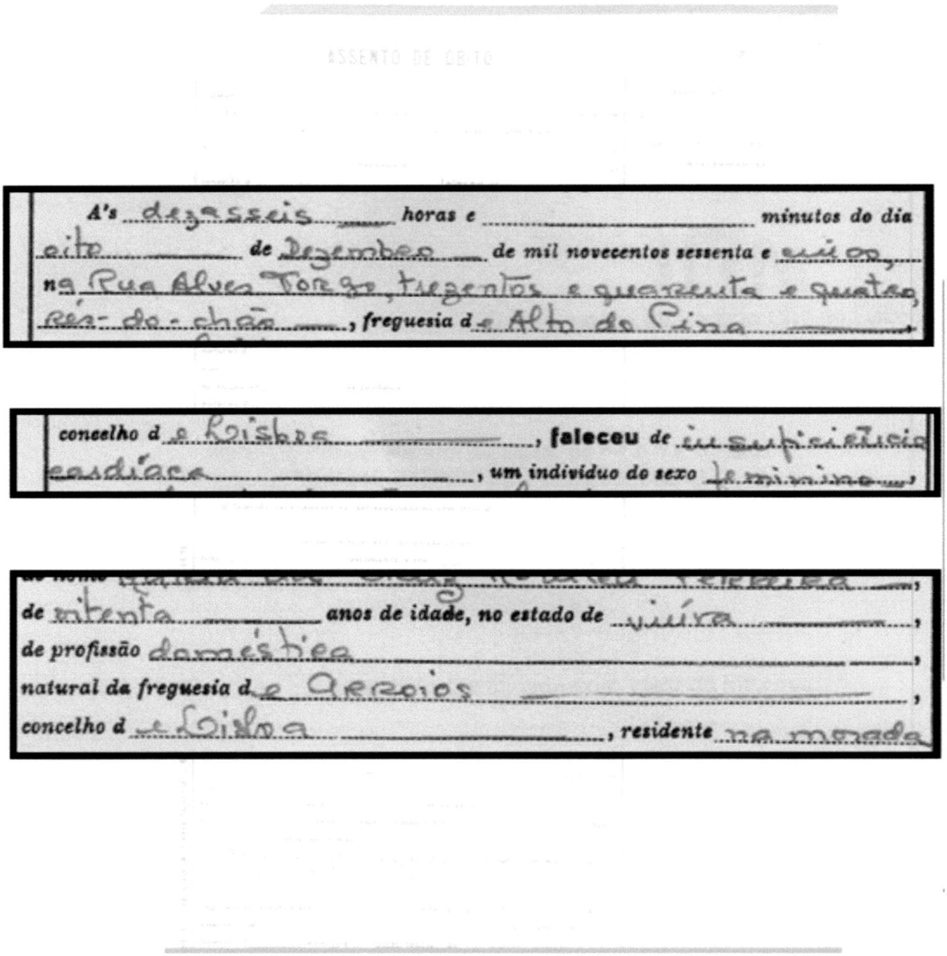
ASSENTO DE ÓBITO

A's dezasseis horas e ... minutos do dia oito de Dezembro de mil novecentos sessenta e cinco, na Rua Alves Torgo, trezentos e quarenta e quatro, rés-do-chão, freguesia de Alto do Pina

concelho de Lisboa, faleceu de insuficiência cardíaca, um indivíduo do sexo feminino,

de oitenta anos de idade, no estado de viúva, de profissão doméstica, natural da freguesia de Arroios concelho de Lisboa, residente na morada

FIGURE 1. EXAMPLE OF BIOGRAPHICAL DATA ASSOCIATED WITH INDIVIDUALS IN A PORTUGUESE IDENTIFIED COLLECTION. A DEATH CERTIFICATE (ASSENTO DE ÓBITO) ASSOCIATED WITH SKELETON #671. THE INFORMATION INCLUDES THE TIME AND YEAR OF DEATH – 1965; ADDRESS AND LOCATION OF DEATH; CAUSE OF DEATH – INSUFICIÊNCIA CARDÍACA (CARDIAC INSUFFICIENCY/HEART FAILURE); SEX – FEMALE; AGE AT DEATH – 80 YEARS OLD; MARITAL STATUS – WIDOW; OCCUPATION – DOMÉSTICA (HOUSEKEEPER).

in Alemán *et al.* 2012; Cardoso 2006; Chi-Keb *et al.* 2013; Cunha *et al.* 2007; Eliopoulos *et al.* 2007; Ferreira *et al.* 2014; Hens *et al.* 2008; Hunt and Albanese 2005; Komar and Grivas 2008; Rocha 1995; Salceda *et al.* 2012; Sanabria-Medina *et al.* 2016). This is exemplified in a Portuguese context in Figure 1.

The biographical data associated with these collections is one of their major assets, and the reason why they have become widely used in anthropological and forensic studies. These enable for the possibility of developing and testing hypothesis-driven research on sex and age at death estimation methods (as discussed in other chapters in this book), bone morphology and population ancestry and variability (discussed by Albanese in this book), as well as to explore bone lesions' correlation with specific diseases and their diagnostic criteria (for example see Cardoso and Henderson 2010; Cardoso and Henderson 2013; Alves-Cardoso *et al.* 2015; Matos and Santos 2006; Milella *et al.* 2014; Perréard-Lopreno *et al.* 2013). Equally, identified skeletal collections also act as proxies to infer an individual's life history, and consequently that of a population. Although fragmented, and despite mediated by dried bones, one of the reasons why identified skeletal collections are so important in anthropological studies (now and in the past) is the fact that the individuals that compose these collections have sometimes been perceived as representative of the population to which they belong. This representation is not limited to the chronological or geographical context, but it also includes the biological, social and cultural contexts in which they existed.

However, many identified skeletal collections have limitations which have been overlooked. Such inherent limitations or biases should be acknowledged and openly discussed, since they have a significant impact on the reconstruction of health and behaviour in past populations. Some are intimately associated with the biographical data of the collection, since historical archives will reflect social and cultural constructs as defined by their inherent time or epoch; others are related to the methodological limitations of inferring sex, age and diseases based on biological tissue, therefore the scope of limitations associated with identified skeletal collections is variable. For example, most identified skeletal collections have a skewed age at death distribution towards older individuals: as these are over-represented within these collections. Further, some collections will contain individuals with a self-reported age at death, as well as ancestral affinity and ethnicity (Komar and Grivas 2008). These are both problematic when developing methods aiming to infer age at death and ancestry estimations, or when exploring diseases that have a strong biological age association, such as degenerative joint diseases (Alves-Cardoso 2008; Jurmain 1999; Weiss and Jurmain 2007). Additionally, social and economic biases may also exist, as identified skeletal collections are described as composed of individuals belonging to the lower social and economic classes, a description overtly applied to Portuguese identified skeletal collections (Cardoso 2006; Cunha and Wasterlain 2007). This classification is based on the occupations of the individuals as listed at death, and on the fact that Portuguese identified skeletal collections are composed of skeletons of individuals declared as *abandoned* by their relatives (Cardoso 2006). The categorization of individuals as *abandoned* is based on cemetery regulation (Law Decree LD411/98). After the period stipulated by law, currently 3 years (note that this period for burial in a temporary grave has been changing over time) human remains may be exhumed from their graves if unclaimed by relatives (Cardoso 2006: 45-46). The fact that remains are considered abandoned has a strong negative social connotation which was and is wrongly viewed by many as being associated with a low social economic and educational status of the individuals (Cardoso and Henderson 2013; Cardoso 2006; Cunha and Wasterlain 2007; Komar and Grivas 2008; Perréard-Lopreno *et al.* 2013; Rocha 1995).

As the source of the remains may introduce bias, biographical data is also in itself a source of bias. Listed occupations at death may be reported by family members, rather than being based on documents from the individual that has died. The occupation name is also ambiguous as to the activities undertaken and therefore represents an oversimplification of tasks performed, and their biomechanical implications. Plus, one must never forget that they are also representative of the historical context in which the individuals lived. Moreover, the manner a researcher decides to use the information is also capable of biasing results and interpretations (Cardoso and Henderson 2013; Milella *et al.* 2015; Perréard-Lopreno *et al.* 2013). To give an example, when exploring to what extent osseous changes and activity were associated in research studies undertaken in identified skeletal collections Perréard-Lopreno and colleagues (2013) highlighted the inconsistency of the interpretative criteria of classification of an occupation, of assignment of a specific occupation to occupational categories, as well as the heterogeneous classification of occupation used. Their research revealed two major points: the fact that *occupation* is an ambiguous historical term, as is its interpretation; and that the latter is highly influenced by the researcher's or study's perspective on the criteria used to classify occupations. Occupations were either classified in accordance with socio-cultural or perceived biomechanical criteria. A socio-cultural criterion was not widely used in past population studies (Perréard-Lopreno *et la.* 2012). The majority of studies chose a biomechanical criterion, classifying occupation on the basis of an association with biomechanical strain due to activity (examples of this can be found in Milella *et al.* 2012; Niinimäki and Baiges-Sotos 2013; Villotte *et al.* 2010).

Despite all of the issues raised above, identified skeletal collections continue to be a major research *tool* in anthropological studies, ranging from past to present population studies, including bioarchaeology, physical anthropology, biological anthropology and forensic anthropology.

Working Life in Bones: Do Skeletons Tell it All?

One of the major areas of research that has benefited from identified skeletal collections is the one dedicated to the study of the reconstruction of past behaviour and activity-related osseous changes (Cardoso and Henderson 2013; Alves-Cardoso *et al.* 2015; Cardoso and Henderson 2010; Jurmain *et al.* 2012; Milella *et al.* 2015; Perréard-Lopreno *et al.* 2013). Activity is inferred from skeletons via the observation of bone changes that have previously been credited as occurring in association with activity. These changes have in the past been referred to as 'Markers of Occupational Stress' (MOS) with the word 'stress' emphasizing biomechanical efforts associated with activity that leaves its mark on the bones and teeth (Jurmain 1999). These marks are considered to be easily identified, and they tend to be mostly associated with joints and muscle attachment sites on the skeleton which are responsible for movement and are therefore believed to possess a strong association with activity (Jurmain 1999). These marks on bones, i.e. activity-related osseous changes have included entheseal changes (formerly referred to as musculoskeletal stress makers or enthesopathies) and changes to the joints. The latter, changes to the joints, have been mostly described in association with degenerative joint disease, with particular reference to osteoarthritis (Alves-Cardoso 2008; Jurmain 1999; Rogers *et al.* 1987; Rogers and Waldron 1995). Other MOS used to assess activity or activity-patterns have included trauma (or specific) fractures, structural adaptations of the bone and functional morphological variations (non-metric traits) (Capasso *et al.* 1998; Jurmain *et al.* 2012; Jurmain 1999; Kennedy 1989, 1998; Larsen 1997). Examples of a traumatic cause of these changes include avulsion fractures, which can be roughly described as the tearing away of a tendon or ligament (including a bone fragment from the attachment site) from the bone due to repetitive micro trauma (Jurmain 1999). This may happen at the ankle, knee, elbow and other major joints of the skeletons due to severe acute stress. Spondylolysis is a slightly different bone change in the sense that although cited as an activity-related pathological lesion (Fibiger and Knüsel 2005), it is also described as a congenital deficiency of bone in the arch of the 5th, or 4th lumbar vertebra. This causes the posterior portion of the vertebra, the pars interarticularis, to separate from the rest of the vertebra, causing localized back pain.

In general, these markers of occupational stress have in common the fact they have been used to infer activity, however they are also multifactorial in origin and therefore MOS may be present in a skeleton and have little to no relationship with activity. Furthermore, limbs and joints are used in a wide range of activities, either related with occupations, hobbies or other uses of the body and it is difficult to allocate any particular change to any specific activity (Jurmain 1999). This observation is pertinent in contexts of bioarchaeological studies. In bioarchaeological studies most of the information on occupation has been determined indirectly using ethnographical accounts, historical data or through exploring the archaeological contexts in which human remains are found. In these cases activity-related osseous changes have mostly been used to explore patterns of activity and occupation associated with major shifts in subsistence patterns, as well as social inequalities and sexual divisions of labour. Identified skeletal collections enable the possibility to directly relate bone changes (e.g. entheseal changes or joint changes) to activities, allowing associations between specific activities and activity patterns to be explored or tested. Entheseal changes consist of three main categories of changes at muscle (as well as tendon and ligament) attachment sites, which include mineralized tissue formation, surface discontinuity and complete loss of the original morphology of the enthesis (for details on the terminology associated with entheseal changes, abbreviated to EC, see Villotte *et al.* 2016). Joint changes observable in dry bone include marginal osteophytes, eburnation, sclerosis and cysts around the articular surface, pitting of joint surfaces and, in severe cases, changes to the joint morphology (Rogers and Waldron 1995; Rogers *et al.* 1987).

Without doubt identified skeletal collections have added an additional layer to the interpretation and reconstruction of behaviour in the past, by using human skeletonized remains, i.e bones, as primary sources of evidence. Identified skeletal collections enable researchers to bring a 'biographical' and 'social and cultural' viewpoint to dried bones. This added perspective is achieved when biographical data includes information on the occupation at death. This information can be used to control for physical effort, biomechanical strain and activity-related changes observed in skeletons, in a similar manner to which sex and age at death listed in the biographical data are also used to test ageing and sexing methods (see Chapters 6 and 7). This chapter specifically focusses on the Portuguese identified skeletal collections from Coimbra University, and the Luis Lopes skeletal collection of the Lisbon Museum of Natural History. Both collections' are associated with the history of teaching anthropology in Portugal, and go back to the 19th century although, in the case of the Lisbon collection, human remains continued to be integrated into the collection until the late 20th, to early 21st century. The skeletons that constitute both collections were retrieved from modern cemeteries (still in use) of the respective cities (Coimbra and Lisbon), with permission of their City Councils. The Coimbra collection was assembled by Eusébio Tamagnini during the years of 1931 and 1942, and it comprises 505 individuals who died between 1802 and 1938 (Alves-Cardoso 2008; Rocha 1995). The Lisbon collection remains have been collected mostly since 1980, and the collection is currently composed of more than 1700 individuals (personal comment by curator). Although a significant part of the collection information still needs to be processed, it is safe to state that the majority of the individuals died between the years 1880 and 1975 (Cardoso 2006).

The definition of what an *occupation* is (or was in the past) is not the major issue when conducting activity-related research on identified skeletal collections. The occupation, as listed at death, is mostly interpreted as static, and consequently well defined in terms of overall tasks and activities performed, so much so that individuals tend to be grouped into overarching occupational categories (Cardoso and Henderson 2013 Villotte *et al.* 2010). The 'occupational categories' used vary according to authors: some use the original data from the archives, referring to the written occupation, i.e. carpenter, shoemaker, police officer, among others (Cardoso and Henderson 2010; Milella *et al.* 2012, 2014); whilst other authors used more inclusive categories, based on the more or less strenuous activities practiced by the individuals, i.e. grouping them into 'manual' or 'non-manual' individuals (Cardoso and Henderson 2013; Cardoso and Henderson 2010; Henderson *et al.* 2013b; Milella *et al.* 2012, 2014; Villotte *et al.* 2010). Independently (or not) of how the information of occupation is manipulated to create occupational categories, it is always employed as a proxy for occupation-specific tasks: activity and work as described in the literature, and its correspondent social and economic status. Consequently, and following this line of reasoning, what is observed in the skeleton and any interpretation based on that observation is an oversimplification of the cumulative impact of all occupational-related activities performed by an individual. It also excludes a significant array of other events and activities which may also have impacted on the skeleton, such as hobbies, changes in occupations and many more daily and mundane activities (Alves-Cardoso 2008; Henderson *et la.* 2013).

Research conducted so far on the topic of entheseal changes and activities has highlighted some issues worthy of consideration, which may have a negative impact on the belief of a simplistic association between biology and 'social and cultural' parameters. For example, Alves-Cardoso's (2008) research has shown that occupations that were historically associated with strong or strenuous activity had, in some cases, lower values of bone changes, when compared to historically less strenuous activities. Furthermore, age at death is a major confounding variable in the analysis of degenerative lesions, both in joints as in entheses (Alves-Cardoso 2008; Cardoso and Henderson 2010). Therefore, although Identified skeletal collections were seen as the perfect setting to test the relationship between bone changes, whilst controlling for age and sex, and at the same time considering the occupation of the individuals, an ultimate interpretation of behaviour still needs to proceed with caution.

Recent studies based on skeletal remains from identified skeletal collections have given rise to a more reflexive approach to the study of bones and skeletons, and their use as proxies for human behaviour and the reconstruction of the past. It has become clear that an individual life history exists beyond biographical data recorded at time of death: it lacks in-depth information of aspects related to a person's life and this has repercussions when trying to infer activity-related bone changes. As stated above, the occupation, as listed at he time of death does not account for all that an individual has performed during his or her lifetime; it does not allow for a glimpse into the more private sphere of a person, or enable the exploration of his or her hobbies and other daily activities that may have affected the musculoskeletal system of which joints and muscle are primary components. Henderson and colleagues (2013) successfully illustrated that in the 19th century rural England activities undertaken by individuals were not stable, varied everyday and could be infrequent. This research was undertaken in identified individuals exhumed from the churchyard of St. Michael and St. Lawrence, Fewston, North Yorkshire, England. The biographical data associated with the individuals allowed access to census records with information on their occupations through time. Furthermore, this research also used published sections of diaries associated with two individuals, John and Mary Dickinson, which refer to activities undertaken by both individuals. The diaries were kept by John Dickinson, their son, between the years of 1878 until his death in 1912, and were later summarised by Harker (1988 *in* Henderson *et al.* 2013a). Consequently, the use of occupation as listed at death lacks detailed information on the frequency and range of activities performed (Alves-Cardoso 2008; Henderson *et al.* 2013a). This has significant relevance in studies aiming to infer activity-related bone changes, as it hinders an evaluation of the impact of the sum of years of activity, or the lack of it, in an individual.

Apart from the issues raised and related to the biological aspect of inferring activity or activity patterns from bone tissue, there is an additional complication related with the social and cultural description of occupation. The term *occupation* is ambiguous if one aims to properly assess specific tasks preformed, their intensity and persistence through life course. Furthermore, one also needs to consider the unclear boundaries of occupational classes, an argument which was made very clear by Alves-Cardoso (2008). The study originally comprised the analysis of a total of 603 individuals, from the Coimbra identified skeletal collection and the Lisbon Luis Lopes skeletal collection.

Of the 603 individuals studied, 299 were from the Coimbra collections (149 males and 150 females) and 304 skeletons from the Lisbon collection (151 males and 153 females). Samples overlapped in time for births and deaths, as individuals were born between 1822 and 1935, and died between 1891 and 1965. Alves-Cardoso research aimed at portraying gender in the Portuguese population of the 19th and early to mid 20th centuries by analysing several markers of occupational stress (MOS), namely entheseal changes and joint changes, and correlating those finds with occupation as listed at death. The assumption underlying this research was that the MOS would reflect the sexual division of labour present in society, and that this would mirror gender constructs. The differential markers of occupational stress would be related to male or female task performance differences. However, as the research developed it was clear that the focus on occupation as listed at death, as a proxy for activity, was a simplification of the complexity associated with each occupation not only in relation to tasks performed, but also in association with the description of the occupation itself. For example, the term 'shop assistant' would be used to classify individuals working not only in small groceries, but also in a fashionable patisserie. The work performed would be similar, but the social status would be profoundly different. The social position of an individual would not only be based on the work performed, but also on access to resources, which ultimately would shape their overall social and biological condition. Other activities of complex categorical description found in official records of the Portuguese identified skeletal collections included those of *proprietário* (male proprietor), or *proprietária* (female proprietor), the English translation being 'proprietor' but of what is unspecified; *empregados de comércio* (shop assistant) and *comerciante* (shop owner) are also nonspecific in the sense that they may be affiliated with a wide variety of occupational work and activities (Alves-Cardoso 2008). However, one of the major limitations is found amongst women, who tend to be classified as *domésticas:* the English equivalent term would be housekeepers. This is an abstract term, in the sense that it provides no solid information as to how classify these individuals in relation to tasks performed: although specific to women, it is overarching in the array of situations it describes ranging from women working in their homes, in the fields as well as 'stay-at-home' women with limited work load who had servants to do the housework. Alves-Cardoso (2008) exposed this bias in the Portuguese identified skeletal collections of Lisbon and Coimbra. The majority of women were (91.1% –276/303) described, under occupation (at death), as being *domésticas* (Figure 2). The remaining women had been listed as servants (the second most frequent female occupation: 5.6% – 17/303, 16 of whom were from the Coimbra sample), dressmakers, teachers and farmers (see Figure 2 for details).

Overall, the sample was composed of older women with an age at death above 50, with individuals' age ranging from 20 to 98 years of age at death (Table 1). Alves-Cardoso (2008) findings showed that changes, with an emphasis on osteoarthritis and entheseal changes, were significantly associated with age. This research was paramount in showing that any joint change used in past population studies as a proxy for occupation was most probably age dependent. Also importantly, it showed that biographical data at death were not without considerable limitations. This is particularly emphasised on the limitations of classifying occupations at death: an even greater problem for women.

Consequently, identified skeletal collections are at the fringe of being the 'ideal' scenario when reconstructing individual behaviour in past societies, despite the fact that they allow for a significant amount of biographical information to be included in the analytic process.

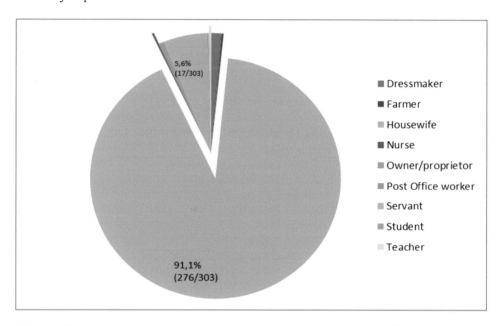

FIGURE 2. DISTRIBUTION OF WOMEN ACCORDING TO OCCUPATION LISTED AT DEATH. THESE RESULTS ARE BASED ON A SAMPLE OF 303 WOMEN, FROM THE COIMBRA IDENTIFIED SKELETAL COLLECTION (N=150) AND FROM THE LISBON LUIS LOPES SKELETAL COLLECTION (N=153) (FOR DETAILS SEE ALVES-CARDOSO 2008).

Occupation	N	Mean	Std. Deviation	Median	Minimum	Maximum
Dressmaker	4	29	8,206	27,5	21	40
Farmer	1	29	.	-	-	-
Housewife	276	56,74	19,594	57	20	98
Nurse	1	72	.	-	-	-
Owner/Proprietor	1	88	.	-	-	-
Post Office Worker	1	73	.	-	-	-
Servant	17	47,12	21,068	43	21	80
Student	1	34	.	-	-	-
Teacher	1	81	.	-	-	-
Total	303	55,96	19,991	56	20	98

TABLE 1. WOMEN'S AGE AT DEATH DESCRIPTIVE STATISTICS BY OCCUPATION AND TOTAL SAMPLE.

What does it mean to be a *Doméstica*?

This section describes a specific case study of one occupation, *domésticas*, to 1) assess the intricate social and cultural understanding of the term; 2) understand the complexity of biomechanical efforts associated with the tasks performed by women that fall, and see themselves, under the classification of *domésticas;* 3) and to see if this is liable of being translated into static terms, such as those composing the biographical information over the collections. The term *doméstica* (housekeeper) is complex and varied because it describes a number of activities (as described above) and tasks undertaken by participants who are exclusively women.

To accomplish this, four women were interviewed, and information on another 3 was collected about female siblings from one of the interviewees. These interviews provide a first-hand approach to determining the meaning of the term '*doméstica*' , incorporating tasks performed, the biomechanical effort attached to these tasks, all life course events of the women (while self-identifying as housekeepers), as well as other activities undertaken (not necessarily because they were housekeepers), simply because they were functional individuals in society. The age range of the participants varied from 65 to 79 years of age (at the time of the interview). Consequently, these women would have been born between 1935 and 1949. This timeframe fits that of the Lisbon and Coimbra identified collections which have been widely used to develop and test methods for identifying activity. This is important since it enables comparison between occupation as listed in the biographical data of the identified skeletal collections, and the account provided by the interviewees.

The interviews focused on: 1) life history – with information on household composition and parents' occupations, as well as age at which work began, change in occupation(s) and information on general health problems that may be occupation-related, as well as other significant pain complaints that may have impacted on daily chores; 2) activity undertaken – occupations undertaken and tasks performed; 3) the amount of time (and duration in years) each task was performed; and 4) association between the identification of a person as *doméstica*, i.e. a woman dedicated mostly to household chores and tasks undertaken. Interviewees were volunteers, and they were known to the author, and were invited to participate in this study following an informal conversation on their life history. The volunteers were fully informed as to the research aims. None agreed to be audio or visually recorded, so information was based on a written interview, anonymity being guaranteed to all participants.

The detailed information about the life history of the interviewees may be consulted in Table 2. All participants describe themselves as *domésticas* at the time of interview. However, when discoursing about their life history they have also used the word *jornaleira* (daily working farmer), *trabalhadora* (worker), *servente* (hired help), as well as *empregada doméstica* (domestic employee or housekeeper) to describe the various occupations and activities they have been engaged in. These latter two classifications were used when describing jobs associated with the household, although they were working as employees for others. However, they also stated that during that period they were simultaneously undertaking domestic tasks at home. All women changed occupation during their lifetime several times, and were consequently exposed to several patterns of biomechanical strain. They were engaged in demanding paid activities starting at an early age, between age 8 to 10. However, one interviewee reported not remembering when she started working (i.e. paid work). The breakdown of activities undertaken by one of the participants (identified as B, 65 years old), the duration through time and the duration of tasks by day may be consulted in Table 3. Overall, this life account is representative of the wide range of activities undertaken by the other women interviewed, as well as the other females in the household (female siblings, and other female relatives). It is noteworthy that the interviewee used the word *doméstica* to refer to a wide variety of activities, performed either within her household as well as while working in paid employment.

It is also interesting to note that when referring to tasks executed while farming they were, in fact, performing activities similar to those being performed by men (as reported by the interviewees). Some of the examples include planting and harvesting crops such as corn, potatoes, rice, grapes, as well as the processing of grapes for wine production, amongst many other tasks.

All participants identified several health issues by their anatomical location, i.e. upper back, lower back, shoulders, elbows, hands, knees and ankles (Figure 3). They also used the term 'spur' to describe the pain felt on the calcaneus (heel). They also referred to traumatic events which affected the wrist. The pain felt by anatomical region was often associated with 'pain from work' (*dores do ofício*): a clear reference to their perceived association with a hard-working life. However, the traumatic incidents were not related to hard work, or with any activity undertaken as work. In both cases the traumatic lesions were accidental, and the result of a fall in both cases whilst going inside the house. Both were also described as 'silly' incidents by the interviewees.

Interviewee	Parents (worked as farmers or related activities)	Number of siblings	Age they started to work	Changed occupation over time	Occupational related health problems
A 78y.o.	✓	3	"as long as can remember"	✓	✓
B 65y.o.	✓	3	10 (paid work)	✓	✓
C 69y.o.	✓	3	8 (paid work)	✓	✓
D 79y.o.	✓	-	12 (paid work)	✓	✓

TABLE 2. GENERAL INFORMATION ON THE LIFE HISTORY OF THE INTERVIEWEES (IDENTIFIED AS A, B, C, D) AND AGE AT TIME OF INTERVIEW.

Activities undertook	Durability of tasks years	Duration of tasks per day	Association between classification and tasks
Worker at rice plantation: plant the rice; collect the rice; transport the rice	10 – 18/23 years old	"Sunrise" to "sunset"	Some tasks had degree of specificities
Shepherded: looking after the animals – walking		Mornings, early afternoon	Some tasks had degree of specificities, which is not conveyed by the classification given
Doméstica: household chores – cleaning the house, washing cloths and carrying it from and to the river, taking care of cattle and picking up their food from the fields and carrying it home	10 – 12 years old (brief period)	Afternoons and evenings	
Jornaleira / day-to-day rural worker: working the land to plant any kind of vegetables, participation in crop plantation and harvesting, vineyard plantation and processing of grapes for wine production, and other farming related activities	23 – until retirement	All day, all year	
Doméstica / day-to-day worker: household chores		When necessary: prepare food; cleaning the house, washing clothes, feeding the animals, help husband in their fields	Same as above
Doméstica: herself			
Doméstica: working the land (subsistence farmer) and taking care of her household	Presently	Same as above in relation to household work: only works in the land to grow vegetables to eat and raise chickens	Same as above
Doméstica: working for others cleaning their houses, and as company lady			

TABLE 3. DETAILED DESCRIPTION OF THE ACTIVITIES UNDERTAKEN THROUGHOUT THE YEARS, WITH ASSOCIATED TASKS AND THEIR CLASSIFICATION FOR INDIVIDUAL IDENTIFIED AS B – WOMAN WITH 65 YEARS OF AGE AT TIME OF INTERVIEW.

Subject	Upper back	Lower back	Shoulders	Elbow	Wrists	Hands	Hip	Knee	Ankle	"Spur"	Trauma
A 78y.o.	✓	✓	right				right				
B 65y.o.	✓	✓	right	right	both	both			both	both	
C 69y.o.	✓	✓			both	both	both	both		both	✓
D 79y.o.	✓	✓				both		both			✓

FIGURE 3. FEEDBACK OF INTERVIEWEES IDENTIFYING THE ANATOMICAL LOCATION IN WHICH 'PAIN' WAS FELT.

These interviews demonstrate the major limitations of the biographical data associated with identified skeletal collections: i.e. that the information at death is not representative of a life history, and therefore should be used with caution in conjunction with the interpretation of bone lesions in the skeletons. This is reinforced in the case of the analysis of non-specific bone changes, or those of multifactorial origin such as entheseal changes (EC), as well as joint changes. Such changes have been widely used to reconstruct past populations' activities, from specific to more general and overarching levels of activities, both within and between populations. Being a *doméstica*, from a biological and biomechanical perspective, is not as simple as grouping all those women together in order to explore so called activity-related bone changes such as EC of joint changes. The amount and variability of work undertaken from an early age until late adulthood (and continued), the varied duration of the tasks undertaken, the specificity of some and the lack of precision of others, are much too complex to enable the development of research designs accurate enough to account for all the amount of work done. To transfer these to the reconstruction of past behaviour is an even more challenging task. All biological considerations aside, being a *doméstica* is also complex from a social and cultural standpoint. Therefore, and in line with what has already been discussed in other published research, the current results show that not all that one experiences is visible in bones or in archival data. Also that not all bone changes can be interpreted as a result of a specific activity. Ultimately, and although identified skeletal collections have paved the way for the construction of research models that allow variables such as sex and age to be controlled for (see above for their importance), a margin for interpretative error continues to exist. In the specific case of occupations with ambiguous meanings this is extremely problematic.

This chapter highlights the importance of an ethnographical approach for a better understanding of the classification of *doméstica*, as self-reported by women themselves; and serves as a model for exploring the diversity of activities and tasks undertaken under the heading of any other occupation. It also allowed to understand the perception women have of the concept of 'hard working life as a *doméstica*' and the pain felt throughout their bodies to be incorporated, creating a direct association between pain and work. This could provide insightful information when exploring occupational health diseases, and how these are understood, incorporated and dealt with in modern days. It also has applicability to the more distant past. Moreover, it made it possible to better understand how these women were using their bodies in work: which specific body movements were done in which tasks, their duration through time (over a day, a week, a month, years), as well as how these may relate with other activities, and with those of men. However, limitations still exist as ethnographical accounts of work, activities, and occupational specific tasks will always be time framed, and inferring back into the past will always be limited to the interviewees age, and their experience on the matter. Nevertheless, and all things considered, this may be a new path enabling the exploration of the complexity of some of the occupations listed in the records of identified skeletal collections.

Concluding remarks

Lives are hardly written in bones. Some events, such as trauma, infections, and chronic diseases may leave a testimony of their passage, but they just leave a hint as to their presence. The 'why' and the 'when' are not written in bones and discovering the answers to these questions is only achievable after careful interpretation of the skeleton and its context. Even in identified collections when 'everything' is known about an individual skeleton there is still missing information. Consequently, interpretations will have an inherent bias. Biographical data that is associated with the individuals of identified skeletal collections are also static information. They represent data about specific moments of a person's life: when he or she was born, and where, where death occurred, what was its cause (as listed in the death certificate); what was their address and occupation. None of this information is without bias, and it is time-specific, in contrast with the fluidity of human life course histories and their many singular events (such as a fall leading to a fracture). Nevertheless, an alternative to explore how individuals used their bodies throughout life can be achieved if other methodologies of assessing occupation are used. The incorporation of self-reported accounts of four women and their working life is illustrative of that.

Acknowledgements

Francisca Alves Cardoso research is funded by the Fundação para a Ciência e Tecnologia (FCT) Investigator Program also supported by the European Commission ESF and POPH [FCT Investigador IF/00127/2014 & FCT Investigador Exploratory Project IF/00127/2014/CP1233/CT0003].

The author would like to than the kindness and willingness of the interviewees to take part on this exploratory project.

References

Alemán, I., Irurita, J., Valencia, A. R., Martínez, A., López-Lázaro, S., Viciano, J. and Botella, M. C. 2012. Brief communication: the Granada osteological collection of identified infants and young children. *American Journal of Physical Anthropology* 149 (4): 606–610.

Alves-Cardoso, F. 2008. *A Portrait of Gender in Two 19th and 20th Century Portuguese Populations: A Palaeopathological Perspective.* Unpublished PhD thesis, University of Durham.

Alves-Cardoso, F., Assis, S. and Henderson, C. 2015. Exploring poverty: skeletal biology and documentary evidence in 19(th)-20(th) century Portugal. *Annals of Human Biology* 43: 102–106.

Buikstra, J. E. and Ubelaker, D. H. 1994. *Standards for Data Collection from Human Skeletal Remains.* Fayetteville, Arkansas Archaeological Survey.

Capasso, L., Kennedy K. A. R. and Wilczak, C. A. 1998. *Atlas of Occupational Markers on Human Remains.* Teramo, Edigrafital S.P.A.

Cardoso, F. A. and Henderson, C. Y. 2010. Enthesopathy formation in the humerus: data from known age-at-death and known occupation skeletal collections. *American Journal of Physical Anthropology* 141 (4): 550–560.

Cardoso, F. A. and Henderson, C. 2013. The categorisation of occupation in identified skeletal collections: a source of bias? *International Journal of Osteoarchaeology* 23 (2): 186–196.

Cardoso, H. F. V. 2006. Brief communication: the collection of identified human skeletons housed at the Bocage Museum (National Museum of Natural History), Lisbon, Portugal. *American Journal of Physical Anthropology* 129 (2): 173–176.

Chi-Keb, J. R., Albertos-González, V. M., Ortega-Munoz, A. and Tiesler, V. G. 2013. A new reference collection of documented human skeletons from Mérida, Yucatan, Mexico. *HOMO – Journal of Comparative Human Biology* 64 (5): 366–376.

Cox, M. 1996. *Life and Death in Spitalfields 1700 to 1850.* York, Council for British Archaeology.

Cunha, E. and Wasterlain, S. 2007. The Coimbra identified osteological collections. Skeletal series and their socio-economic context. *Documenta Archaeobiologiae* 5: 23–33.

Eliopoulos, C., Lagia, A. and Manolis, S. 2007. A modern, documented human skeletal collection from Greece. *HOMO – Journal of Comparative Human Biology* 58 (3): 221–228.

Fibiger, L. and Knüsel, C. J. 2005. Prevalence rates of spondylolysis in British skeletal populations. *International Journal of Osteoarchaeology* 15(3): 164–174.

Ferreira, M. T., Vicente, R., Navega, D., Gonçalves, D., Curate, F. and Cunha, E. 2014. A new forensic collection housed at the University of Coimbra, Portugal: the 21st century identified skeletal collection. *Forensic Science International* 245: 202.e1–202.e5.

Henderson, C. Y., Craps, D. D., Caffell, A. C., Millard, A. R. and Gowland, R. 2013a. Occupational mobility in 19th century rural England: the interpretation of entheseal changes. *International Journal of Osteoarchaeology* 23 (2): 197–210.

Henderson, C. Y., Mariotti, V., Pany-Kucera, D., Villotte, S. and Wilczak, C. A. 2013b. Recording specific entheseal changes of fibrocartilaginous entheses: initial tests using the Coimbra method. *International Journal of Osteoarchaeology* 23 (2): 152–162.

Henderson, C. Y. and Gallant, A. J. 2005. A simple method of characterising the surface of entheses. Poster presented at the *Thirty-Second Annual Meeting of the Paleopathology Association*, Milwaukee.

Hens, S. M., Rastelli, E. and Belcastro, G. 2008. Age estimation from the human os coxa: a test on a documented Italian collection. *Journal of Forensic Sciences* 53 (5): 1040–1043.

Hunt, D. R. and Albanese, J. 2005. History and demographic composition of the Robert J. Terry anatomical collection. *American Journal of Physical Anthropology* 127 (4): 406–417.

Jurmain, R. 1999. *Stories from the Skeleton. Behavioral Reconstruction in Human Osteology.* Australia, Gordon and Breach Publishers.

Jurmain, R., Alves-Cardoso, F., Henderson, C. and Villotte, S. 2012. Bioarchaeology's Holy Grail: the reconstruction of activity. In A. Grauer (ed.), *A Companion to Paleopathology*: 531–552. Malden, Wiley/Blackwell.

Kennedy, K. A. R. 1989. Skeletal markers of occupational stress. In M. Y. Iscan and K. A. R. Kennedy (eds.), *Reconstruction of Life from the Skeleton*: 129–160. New York, Wiley-Liss.

Kennedy, K. A. R. 1998. Markers of occupational stress: conspectus and prognosis of research. *International Journal of Osteoarchaeology* 8: 305–310.

Komar, D. A. and Grivas, C. 2008. Manufactured populations: what do contemporary reference skeletal collections represent? A comparative study using the Maxwell Museum documented collection. *American Journal of Physical Anthropology* 137 (2): 224–233.

Larsen, C. S. 1997. *Bioarchaeology. Interpreting Behaviour from the Human Skeleton.* Cambridge, Cambridge University Press.

Larsen, C. S. 2002. Bioarchaeology: the lives and lifestyles of past people. *Journal of Archaeological Research* 10: 119–166.

Matos, V. and Santos, A. L. 2006. On the trail of pulmonary tuberculosis based on rib lesions: results from the human identified skeletal collection from the Museu Bocage (Lisbon, Portugal). *American Journal of Physical Anthropology* 130: 190–200.

Milella, M., Alves-Cardoso, F., Assis, S., Perréard-Lopreno, G. and Speith, N. 2014. Exploring the relationship between entheseal changes and physical activity: a multivariate study. *American Journal of Physical Anthropology* 142 (2): 215–223.

Milella, M., Belcastro, M. G., Zollikofer, C. P. E. and Mariotti, V. 2012. The effect of age, sex, and physical activity on entheseal morphology in a contemporary Italian skeletal collection. *American Journal of Physical Anthropology* 148 (3): 379–388.

Molleson, T. and Cox, M. 1993. *The Spitalfields Project, volume II – The middling sort.* Report 86. York, Council for British Archaeology Research.

Niinimäki, S. and Baiges-Sotos, L. 2013. The relationship between intensity of physical activity and entheseal changes on the lower limb. *International Journal of Osteoarchaeology* 23 (2): 221–228.

Perréard-Lopreno, G., Alves-Cardoso, F., Assis, S., Milella, M. and Speith, N. 2013. Categorization of occupation in documented skeletal collections: its relevance for the interpretation of activity-related osseous changes. *International Journal of Osteoarchaeology* 23 (2): 175–185.

Rocha, M. A. 1995. Les collections ostéologiques humaines identifiées du Musée Anthropologique de l'Université de Coimbra. *Antropologia Portuguesa* 13: 7–38.

Rogers, J., and Waldron, T. 1995. *A Field Guide to Joint Diseases in Archaeology.* New York, John Wiley.

Rogers, J., Waldron, T., Dieppe, P. and Watt, I. 1987. Arthropathies in paleopathology: the basis of classification according to most probable cause. *Journal of Archaeological Science* 14: 179–193.

Salceda, S. A., Desántolo, B., García-Mancuso, R., Plischuk, M. and Inda, A. M. 2012. The 'Prof, Dr. Rómulo Lambre' collection: an Argentinian sample of modern skeletons. *HOMO – Journal of Comparative Human Biology* 63 (4): 275–281.

Sanabria-Medina, C., González-Colmenares, G., Restrepo, H. O. and Rodríguez, J. M. G. 2016. A contemporary Colombian skeletal reference collection: a resource for the development of population specific standards. *Forensic Science International* 266: 577. e1–577.e4.

Scheuer, J. L. and Black, S. M. 1995. *The St. Bride's Documented Skeletal Collection.* Unpublished archive held at the Biological Anthropology Research Centre, Department of Archaeological Sciences, University of Bradford.

Scheuer, J. L. and Bowman, J. E. 1995. Correlation of documentary and skeletal evidence in the St. Bride's Crypt population. In S. R. Saunders and A. Herring (eds.), *Grave Reflections: Portraying the Past Through Cemetery Studies:* 49–70. Toronto, Canadian Scholars Press.

Villotte, S., Assis, S., Alves-Cardoso, F., Henderson, C. Y., Mariotti, V., Milella, M., Pany-Kucera, D., Speith, N., Wilczak, C. A. and Jurmain, R. 2016. In search of consensus: terminology for entheseal changes (EC). *International Journal of Paleopathology* 13: 49–55.

Villotte, S., Castex, D., Couallier, V., Dutour, O., Knüsel, C. J. and Henry-Gambier, D. 2010. Enthesopathies as occupational stress markers: evidence from the upper limb. *American Journal of Physical Anthropology* 142 (2): 224–234.

Weiss, E. and Jurmain, R. 2007. Osteoarthritis revisited: a contemporary review of aetiology. *International Journal of Osteoarchaeology* 17: 437–450.

Chapter 9

The Fate of Anatomical Collections in the US: Bioanthropological Investigations of Structural Violence

Rachel J. Watkins[1]

[1] American University

Introduction

In the US, the ongoing quest for richer historical and social readings of US-based anatomical collections includes a recent turn toward a structural violence framework. Peace studies scholar Johan Galtung developed the concept to explain the harm done to individuals and groups by way of socially embedded constraints that limit access to resources and power (Galtung 1969). These structural forces play a key role in direct injury or death.

Paul Farmer is credited with proposing an anthropological orientation toward structural violence to understand social inequalities in modern life (Farmer 2004). He argues that understanding the full impact of structural violence on groups of people involves not only conversations with the living (through which 'ethnographic visibility' is achieved), but also the dead (Farmer 2004). Granted, Farmer was speaking to social conditions related to Haitians disproportionately suffering from HIV/AIDS, tuberculosis and other health disparities. Engagement with the dead is necessary for identifying those who succumbed to disease and correctly 'tallying' the body count resulting from the impact of structural inequalities on health (Farmer 2004, 307).

Different contexts aside, bioanthropologists[1] are drawing on this line of thinking to frame the lived *and* post-mortem experiences of individuals in US-based anatomical collections. This is because the acquisition, autopsy and dissection of individuals for teaching and research was informed by their locations on the social and economic margins of the societies in which they lived. Therefore, structural inequalities in part determined whose bodies were acquired for use in medical schools and laboratories. This turn toward a bioanthropology of structural violence represents a necessary step toward the study and analysis of anatomical collections similar to those already taking place in Europe. The volume *The Fate of Anatomical Collections* exemplifies the historical *and* cultural readings of European anatomical collections that place them at the center of investigating interactions between the arts, humanities, sciences and social sciences (Knoeff and Zwijnenberg 2015). I want to propose that a more intentional exploration of these interactions is crucial for appropriately situating US-based anatomical collections in the past and present. This includes integrating

---

[1] The terms biological anthropologists and bioanthropologists will be used throughout the paper interchangeably.

cultural studies scholarship centered on the histories of people represented in these collections (Blakey 2001; Watkins and Muller 2015). There is an American intellectual tradition of situating this scholarship outside the realm of 'mainstream' relevance. However, this is the very literature that is assisting with nuancing understandings of biological variation in these samples by way of ethnicity and social location rather than race (de la Cova, 2012; Pearlstein 2015; Watkins 2015). Under these circumstances, this is an important element of minimizing the perpetuation of structural violence upon individuals in these collections.

Statement of the Problem

US-based anatomical collections continue to be studied in siloes, with research based in different disciplines isolated from one another. Studies produced by historians are often limited to the actors involved in the creation of the collection, and the social and intellectual climate in which they were created. The people in the collections have little to no visibility. For instance, Ann Fabian's *Skull Collectors* (2010) focuses on Samuel Morton as architect of a collection of crania that played an important role for racial science. Redman's *Bone Rooms* (2016) discusses human remains as highly sought after resources for establishing scientific authority during and after anthropology's turn away from racial science. In both cases, the individuals in the collections are discussed primarily as data. (Bio)anthropological studies of anatomical collections tend to focus on the biological status of individuals in the collection - albeit in the context of sociohistorical location. Moreover, this research is characterized as being somewhat segmented from related branches of history and cultural studies. As a result of the history of American racism that permeates research practices and knowledge production, social science, humanistic and activist approaches are not well integrated into bioanthropological research (Blakey 2001; Watkins and Muller 2015).

As a redress, our use of a structural violence framework is leading to studies that situate collections within the history of medicine, science and the social and biological lives of the people in the collections. However, we need to draw more deeply upon the concept of structural violence as both Galtung and Farmer articulate it to maximize the critical and humanistic study of these collections. Michael Blakey coined this term to describe an approach to researching human remains that recognizes the social and political influences on past and present scientific studies (1998). Therefore, in the same way that we tie science of the past to social conventions of the time, we should also consider how our current research practices reflect the same ideological positioning (Blakey 1998; Watkins and Muller 2015). Specifically, we should evaluate the extent to which our current research practices fall inside and outside of the realm of exacting some form of structural violence on the people whose remains we study. Recent commitments to a critical and humanistic study of human biology make the appropriation of this concept unsurprising, and there are several stated *general* admonitions to consider our role in perpetuating structural violence upon the people we study (Lans 2016; Nystrom 2014; Zuckerman et al. 2014). However, we have yet to consider the *specific* forms that structural violence might take on our part. This is one of many ways that we can move beyond examining structural violence in the creation of collections and consider how our research is implicated in shaping the future and *fate* of the collections (Knoeff and Zwijnenberg 2015; Nystrom 2016). Central to the premise of the papers in The *Fate of Anatomical Collections* is that anatomical collections are fluid - the purpose, appearance and meaning change based on the cultural and scientific ideas of their keepers (Knoeff and Zwijnenberg 2015, 5). As they are products of cultural and scientific practices, it is not possible for them to be treated as 'finished objects.' The records associated with these collections are central to their ability to be mobile across time (and in some cases) space. Therefore, the term fate is not used to suggest some final point at which collections end up. Rather, it is used to conceptually place the collections on a continuum of changing meanings and uses.

This paper draws upon the concepts of mobilization and fate to frame changes in the meanings and uses of the W. Montague Cobb skeletal collection and associated documents over time. This includes how the collection was and is articulated with remains, artifacts and other biological data from the New York African Burial Ground (NYABG) project. There is an established history of the collection's use as a tool for creating space for African-Americans to participate in bioanthropological research, medical school training and constructing alternative understandings of racial variation that speak to its political mobilization. As such, the collection is recognized as part of an interdisciplinary movement addressing intellectual and ethical issues related to the biological, social and economic impacts of typological studies (Rankin Hill and Blakey 1994, Watkins 2007; Watkins and Muller 2015). Mobilization will also be discussed in terms of the literal and varied consolidation of skeletal and documentary data sets within the archive by scholars conducting research on the collection.

Past and present mobilizations of skeletal and documentary data point to the importance of documents including personal data and medical history in identifying how and what social data are embedded in the skeletal remains (Higgins et al. 2002; Phillips 2001, 2003). This has important implications for their cross disciplinary relevance as historical texts and material records of social and biological lived experience. However, in the context of the social history of the collection, we also need to consider how these textual 'traces' allow us to chronicle different uses and meanings of the collection. For instance, communication between Cobb and his colleagues about acquiring remains, data collection and analytical frameworks can be used to identify the movement of the collection through different research and professional transitions in the discipline. Ultimately, tracing changes in how the collection is used and articulated with other collections over time provides visibility that prevents exacting structural violence through obscuring history (Farmer 2004).

This paper begins with a brief history of the Cobb skeletal collection, followed by a discussion of various stages of the collection's mobilization. I consider what this movement suggests about how our present day research practices might (un) knowingly tend toward exacting structural violence upon individuals in anatomical collections. I conclude the paper with questions that I argue we should we ask ourselves to fully integrate this framework into our critical and humanistic-oriented research practices.

The Skeletal Collection and Texts

The W. Montague Cobb Human Skeletal Collection was one of several data sets housed in the anthropology laboratory that Cobb established in 1932 after completing doctoral studies in biological anthropology at Western Reserve University (Cobb 1936). Remains acquired prior to the mid-1950s were unclaimed for burial by relatives and in keeping with the laws at the time, were distributed to medical schools for use in gross anatomy classes. The history of the collection as well as details of its composition are chronicled in a number of publications (see Blakey 1988, 1995; de la Cova 2012; Rankin Hill and Blakey 1994; Watkins 2007, 2010, 2012). In general, the collection is described as a sample of predominantly African-American individuals living in Washington DC at the time of death (Blakey 1988, 1995). However, the collection also includes Chinese, East Asian, and European and US-born 'whites'.

 Records including variable amounts of data accompanied each individual, including age, sex, ethnicity, occupation, nativity and cause of death. Some records also include photographs, drawings, measurements and notes on specific medical conditions. Files indicate that most individuals were day laborers or domestics and lived in the Northwest quadrant of the city. However, a substantial number of individuals in the collection were residents of the city's poor house (Watkins 2010). Because of ethnicity and class markers attributed to the sample, the collection is considered to be representative of a socially and economically marginal population.

Cobb continued processing cadavera until 1969.The collection originally consisted of 932 adult skeletons and 48 infant remains. However, this number has been reduced to roughly 650 individuals after a period of long-term storage in the basement of the Howard University School of Medicine after Cobb's retirement in 1974.

Early Mobilization of a Cadaver and Skeletal Collection

A discussion of the full range of research and activities encompassed within the laboratory Cobb established at Howard University falls outside of the scope of this paper. Nonetheless, it is important to note that the development of his skeletal collection cannot be separated from the development of other biological data; for establishing the skeletal collection was part of a larger suite of activities geared toward equipping an African American institution to conduct anatomical, growth and development and other biological studies of African Americans (referred to as Negroes at the time). Cobb believed these data to be essential to countering scientific arguments in favor of racial typology and hierarchy (Cobb 1936; Rankin Hill and Blakey 1994; Watkins 2007, 2012. Therefore, it is important to recognize that the skeletal collection first existed as cadavers used for anatomical instruction. Similarly, the texts now associated with the skeletal collection were amassed during this time, and were first used as part of medical school instruction. All anatomy class schematics included age, sex, 'race' and cause of death (Figure 1).

CAUSES OF DEATH, 1958-59 CADAVERA

TABLE 1
Cad. No. 483
Male-Negro
Age 65
Died 1/15/53
Cause of Death:
Acute cardiac failure
exposure pneumonia

TABLE 2
Cad. No. 625
Male-White
Age 83
Died 5/16/58
Cause of Death:
Congestive heart failure,
arteriosclerotic heart
disease; inoperable
cancer of Prostate

TABLE 3
Cad. No. 604
Male-White
Age 79
Died 9/10/57
Cause of Death:
Acute congestive heart
failure, myocardiosis
chronic

TABLE 4
Cad. No. 608
Male-Negro
Age 57
Died 10/22/57
Cause of Death:
Bronchopneumonia,
Carcinoma of hypopharynx

TABLE 5
Cad. No. 612
Male-Negro
Age 60
Died 12/17/57
Cause of Death:
Acute congestive heart,
failure; arteriosclerotic

TABLE 6
Cad. No. 222
Male-Negro
Age 67
Died 10/12/37
Cause of Death:
Not known

TABLE 7
Cad. No. 618
Male-White
Age 55
Died 11/26/57
Cause of Death:
Acute congestive heart
failure, cardiac asthma
myocardosis chronic

TABLE 8
Cad. No. 000
Male-Negro
Cause of Death:
Bronchogenic carcinoma

TABLE 9
Cad. No. 603
Male-Negro
Age 47
Died 10/5/57
Cause of Death:
Acute myocardial infarction
arteriosclerotic heart disease
generalized arteriosclerosis

TABLE 10
Cad. No. 620
Male-Negro
Age 50
Died 12/31/57
Cause of Death:
Bronchopneumonia

TABLE 11
Cad. No. 599
Male-Negro
Age 59
Died 8/25/57
Cause of Death:
Carcinoma of bladder

TABLE 12
Cad. No. 610
Female-White
Age 93
Died 10/29/57
Cause of Death:
Toxemia and exhaustion.
generalized arteriosclerosis

TABLE 13
Cad. No. 607
Male-Negro
Age 69
Died 10/20/57
Cause of Death:
Acute hepatic insufficiency.
Cirrhosis, hepatoma and
metastatic malignancy

TABLE 14
Cad. No. 598
Female-Negro
Age 75
Died 9/2/57
Cause of Death:
Cerebral thrombosis,
arteriosclerotic cardio
vascular disease. Generalis
ed arteriosclerosis

TABLE 15
Cad. No. 628
Male-Negro
Died 5/12/58
Cause of Death:
Broncho pneumonia, arteri.
sclerotic heart disease.
Generalized arterio
sclerosis

TABLE 16
Cad. No. 630
Male-White
Age 67
Died 5/17/58
Cause of Death:
Acute myocardial infarction
due to arteriosclerotic
heart disease

TABLE 17
Cad. No. 626
Male-Negro
Age 76
Died 5/13/58
Cause of Death:
Chronic congestive heart
failure, arteriosclerotic
c. vascular disease
generalized arteriosclerosis

TABLE 18
Cad. No. 629
Male-Negro
Age 88
Died 5/10/58
Cause of Death:
Cerebral vascular accident
with right hemiplegia,
hemorrhage cystitis

TABLE 19
Cad. No. 624
Male-White
Age 88
Died 3/16/58
Cause of Death:
Congestive heart failure
arteriosclerotic cardio
vascular disease

TABLE 20
Cad. No. 623
Male-White
Age 73
Died 3/22/58
Cause of Death:
Acute cerebral vascular
accident hemorrhage

TABLE 21
Cad. No. 622
Male-White
Age 70
Died 3/18/58
Cause of Death:
Ruptured aortic aneurysm
malnutrition

TABLE 22
Cad. No. 621
Male-Negro
Age 83
Died 3/20/58
Cause of Death:
Bronchopneumonia

TABLE 23
Cad. No. 609
Male-Negro
Age 42
Died 10/25/57
Cause of Death:
Acute pulmonary edema.
Bronchopneumonia

TABLE 24
Cad. No. 631
Female-White
Age 47
Died 5/11/58
Cause of Death:
Prob. Pulmonary embolism,
septicemia sub acute; bacterial
endocarditis

TABLE 25
Cad. No. 627
Male-Negro
Age 49
Died 5/8/58
Cause of Death:
Uremia due to Glomerulo
nephritis

FIGURE 1. SCHEMATIC OF CADAVERA AS ARRANGED IN THE ANATOMY LABORATORY INCLUDING DEMOGRAPHIC INFORMATION.

It was standard practice for students in Cobb's anatomy students to study these records along with the cadaver and (later) the skeleton, as it was essential to understanding how mortality statistics and demographics were informed by social context (Cobb n.d. 1935, 1936). This interface was particularly important in understanding how cadaver populations historically represent particular segments of the population (which Cobb referred to as 'poorly circumstanced').[2] Therefore, the biocultural syntheses and health disparities reflected in such collections is not is not associated with being representative of the general population (Cobb n.d.). This orientation was the cornerstone of Cobb's approach to studying human biological diversity in a way that did not rely upon racial typology to explain it. Early data gathered from the skeletal collection was included in Cobb's article titled, 'The Physical Constitution of the American Negro,' published in the Journal of Negro Education in 1934. Cobb referenced the collection in arguing against the notion of racial differences in body type (Cobb 1934).

The skeletal collection numbered at roughly 550 by the 1950s, during which time systematic data collection to produce survey reports and other descriptive statistics began. A detailed pathological study conducted by Pearl Lockhart Rosser in 1959 is housed in the Cobb laboratory, including observations of fractures, lesions associated with infectious disease, cases of spondylitis, ankylosis and skeletal anomalies. She also calculated the percentages of whites, blacks, males and females in the sample. In terms of mobilizing data, the focus on testing for the statistically significant presence of various conditions in the sample marked a period of greater concentration on skeletal remains, with less attention paid to personal records.

Upon Cobb's retirement from Howard University in 1974, skeletal remains were placed in the basement of the medical school, where they stayed until custodianship was transferred to Professor Michael Blakey in 1987. The curatorial activities and research carried out on the Cobb skeletal collection immediately after being removed from storage are chronicled in Blakey's initial report (1988). This includes establishing an alpha series made up of the most complete skeletons with the most detailed personal records (Blakey 1988, 1995). The subsample provided a reliable data set for conducting statistical analyses and systematic collection of personal information while curation was underway. In addition to constructing an alpha series, Blakey assigned codes ranging from 'A1' to 'F1' to organize skeletons into groups according to the number of elements present and the level of detail in personal records. Therefore, codes indicated the suitability of skeletons for various studies. Curatorial activities and laboratory preparation continued between 1992 and 1994 under a grant from the National Science Foundation. Funds were allocated for the renovation of laboratory space, transferring the remaining skeletons in storage to the newly renovated laboratory after completion and establishing a database (Blakey 1988, 1995).

---

[2] Cobb's article 'Municipal History from Anatomical Records' (1935) was his first key statement on his biocultural approach to studying remains from anatomical collections. This involved taking into consideration 'sociological and historical facts' that indicate that individuals came from the 'least stable elements of marginal economic groups in the living population' (157). The paper was a by-product of his dissertation research on what is now called the Hamann-Todd anatomical collection in Cleveland, OH. The collection he established at Howard was structured to model that collection (Rankin-Hill and Blakey 1994; Watkins, 2007).

Student training in human osteology and bioanthropological research methods was an integral part of developing the laboratory and its collections during the 1990s. A published dental anthropological study co-authored with a student and faculty colleague in The Department of History was an early outcome of this effort (Blakey et al. 1994). Training also led to a number of students declaring majors in anthropology, as well as pursuing graduate work in biological anthropology. This represents yet another phase in which the collection was mobilized to redress the political and intellectual issues related to increasing the presence of African Americans in the field.

Research on skeletons from the New York African Burial Ground (NYABG) also took place in the Cobb Research Laboratory (Blakey and Rankin-Hill 2004). When the curation of the Cobb skeletal collection was nearly complete in 1991, human remains were uncovered on Lower Manhattan property owned by the General Services Administration as a result of a construction project. The site was a former cemetery marked as the 'Negroes Burying Ground' on maps dated to the 1700s (Blakey 1998; Blakey and Rankin-Hill 2004; Laroche and Blakey 1997). As public concern around the site and the treatment of the uncovered remains grew, Blakey emerged as a key figure to helm a research study based at Howard University according to the wishes of the descendant community. Blakey and colleagues drafted an interdisciplinary research design that included skeletal, archaeological, genetic and historical analysis. Students previously trained in the laboratory made up a substantial portion of support personnel.

The high-profile nature of the NYABG Project, including its significance to larger discussions about American history, slavery and the development of colonial New York, somewhat eclipsed the visibility that Blakey attempted to promote for the Cobb skeletal collection in the mid-1990s (Blakey 1995). Nonetheless, the laboratory facilities and personnel that allowed Blakey and researchers to successfully carry out NYABG research at Howard University were made possible by external funding used to establish a permanent space for housing and conducting research on the Cobb collection. This means that the Cobb collection played a role in helping to expand intellectual and ethical discourses involving research conducted on African descendant skeletal remains. The role that these collections play in each other's fate will be discussed in a following section of this chapter.

Twenty-first Century Mobilizations

I note elsewhere that the number of studies using the Cobb skeletal collection – relative to the number of years the collection has been available for research – are small. Initial research conducted by Watkins and Muller (2003 and 2006 respectively) prioritized the identification of individuals and groups within the collection within a historical, social, and biological context that did not privilege statistical analysis. For instance, Watkins focused part of her comparative analysis on individuals who lived in the District's poor house and those who lived in residential neighborhoods. Differences in health between the two groups appear to reflect differences in access to social services and support. However, unlike poor house samples from Upstate New York, people in residential neighborhoods show more infectious disease burden and biological stress related to physical demand (Watkins, 2003, 2012). Historical documents indicate that conditions in the poor house were not ideal, let alone categorically 'healthy.' However, residence in the poor house involved incorporation into the city's relief system. Therefore, the health and disease patterns observed in people living throughout the city reflected their lack of access to these services. Similarly, Muller's study of trauma in the Cobb Collection points to uniquely high degrees and patterns of trauma compared with other skeletal collections. She attributes her findings to the politics of racism that restricted blacks to an occupational status associated with the most menial and physically demanding jobs. Muller's interpretation of the frequency and patterning of fractures and dislocations indicates a high level of trauma associated with accidents and interpersonal violence (Muller, 2006). Remains from the Cobb Collection are included in several of de la Cova's analyses of trauma and other aspects of health in 19th century anatomical collections (de la Cova, 2010, 2011). These studies highlight the multiple social locations of individuals in terms of statistical and sociohistorical analysis. Specifically, she organized samples into multiple cohorts based on birth relative to Reconstruction (the time period immediately after enslavement in the US), ethnicity, status as free or enslaved, as well as combinations of these categories. Statistically significant differences in infectious disease burden between ethnic groups were associated with local and extra-local environmental stressors affecting birth during reconstruction (de la Cova, 2011). de la Cova's efforts toward a cultural analysis of skeletal remains also considered how ethnically-specific ways of performing masculinity were reflected in trauma patterns among males (de la Cova, 2010).

It is also worth noting that de la Cova was the first to draw upon the concept of structural violence in the analysis of an anatomical collection. Her 2012 study of 256 European American and African American females found the latter group had more fractures to the face and hand suggestive of intimate partner violence, while the former group had significantly higher incidences of hip and arm fractures, suggestive of accidental injuries. Documentary analysis allowed her to obtain evidence that both African and Euro-American women in the sample were institutionalized. She concluded that trauma patterns she observed could not be solely understood in terms of interpersonal violence. They are also likely to reflect the interactions between the state and individuals subjected to institutional protocols.

Current Mobilizations: Mitigating Present-Day Structural Violence

A paper titled, 'Dissection and Documented Skeletal Collections: Embodiments of Legalized Inequality' factors the Cobb collection into recent discussions framing anatomy and dissection practices as a form of structural violence (Muller, Pearlstein and de la Cova 2016). The paper is included in an edited volume titled, *The Bioarchaeology of Dissection and Autopsy*, which focuses on the deliberate practices around establishing collections rooted in beliefs about the varied worth of particular individuals and groups (Nystrom 2016). The discussion subjects the collection to yet another form of mobilization by way of articulating its history with that of the Huntington and Hamann Todd collections.

Extending the concept of structural violence past the process of creating the samples brings different historical considerations into view. For instance, twenty-first century studies of the Cobb collection indeed use documents to understand the social context of health and disease patterns observed in the skeletal sample (Blakey et al 1994; de la Cova 2010, 2011; Muller 2006; Watkins 2003, 2012). However, the skeletal remains that are now available for study do not reflect the demographic breadth of the original sample. Census records did not suffer the same fate, as important demographic, medical, etc. information for all individuals in the original sample were maintained. Based upon these circumstances, Watkins and Muller proposed situating the skeletal remains and documents as a collective archive from which to develop a sample that includes individuals for whom we have skeletons *and* individuals for whom skeletons no longer exist (Watkins and Muller 2015). This mobilization of skeletal remains and documents is the result of critical reflection upon how a singular focus on the skeletal remains privileges the scientific investigation of individuals over other forms of analysis. In other words, the focus on existing skeletal remains reflects an investment in producing scientific knowledge rather than knowledge of the population sample. As such, this potentially constitutes a form of structural violence upon individuals in the collection that researchers can exact. Consider Farmer's assertion that the erasure of history is a mechanism and result of structural violence (2004). The archival approach that Watkins and Muller propose allows for a historical treatment of the collection that is not possible if a researcher limits the study of the collection to existing skeletal remains. The inclusion of individuals for whom we no longer have skeletons in studies allows for a study sample that includes infants, older subadults, and named 'whites' who were a part of the original skeletal collection. A sample that is more representative of all age and demographic categories is one that makes for a more comprehensive and inclusive history of the collection and the people within it.

Arguably, there are other aspects of our current research practices that should be considered in the context of structural violence. The people in the Cobb collection, like those in other US-based anatomical collections, are described as 'unclaimed,' were not subject to burial and are housed in the protective environment of a laboratory bearing Cobb's name. As a result, there is a level at which individuals in this and other collections are assumed to be perpetually 'available' for study. Might this also be a form of structural violence? The question is worth raising in light of some differences between how the Cobb and NYABG collections were and are subject to research and public engagement. The differences speak to an interesting condition of research on anatomical collections: bioarchaeological method and theory is applied broadly to the study of anatomical collections. However, the attendant social and political concerns that inform research and treatment are not largely applied. Therefore, the standardized markers of anatomical collection only allow for limited forms of politicization.

In the absence of formal legal protections (such as NAGPRA), the social and political unrest following the rediscovery of the New York African Burial Ground led to a destabilizing of a certain type of racialized research subject. Individual citizens and activists within and outside of New York claimed social, cultural and spiritual kinship with the people interred at the African Burial Ground based on factors such as: 1) a shared African origin rooted in the forced transport to the Americas; and, 2) the shared cultural experience of oppression based on the continued racialization of African descendant people in the US. Based on this sense of connection, these constituents self-identified as a Descendant Community acknowledged by the research team as having a right to have a say in the research design and methods used to study the NYABG remains. For instance, the Descendant Community stressed the importance of studying the remains in both a Diasporic context rather than a history solely defined by enslavement. They also participated in decision making regarding laboratory practices, as well as determining the ultimate resting place for the remains after research was complete. Therefore, the remains from the NYABG were subject to repatriation and reburial, being reinterred after a period of research (Blakey 2004, 2010; Blakey and Rankin-Hill 2004; Kakaliouras 2012).[3]

This is not to say that ethics and standards do not inform [our] work on skeletal remains in anatomical collections. In addition to the standards and guidelines in accordance with our research and professional organizations, there are stringent criteria for gaining access to anatomical collections for study. In the case of the Cobb collection, the submission of a research proposal is required that is reviewed by the Laboratory Director and board of researchers affiliated with the laboratory (Watkins and Muller 2015). Nonetheless, there is not a level of oversight in laboratory research – let alone public engagement and interpretation – that destabilizes the researcher-research subject relationship as it is constructed by the researcher, or interrogating the laboratory space as a natural environment for the collection (Watkins n.d.; 2015).

The structure of laboratory tours during the NYABG project research phase reflect efforts to mitigate this dynamic. Public education and interpretation was a hallmark of the NYABG project research process. Blakey as Scientific Director, Laboratory Director Mark Mack and staff members regularly provided tours to schools, organizations and interested members of the general public. The critical humanistic orientation that informed the study of both the Cobb and NYABG collections led to the Cobb collection being included in NYABG project laboratory tours. Rather than an isolated presentation of the NYABG research, visitors learned that the Cobb laboratory housed the remains of individuals representing over 250 years of biological and social history of African Americans. This included presenting skeletal remains from both collections during tours. This is quite different from the access members of the general public usually have to anatomical collections. Public engagement with the Samuel Morton collection at the University of Pennsylvania also stands as a rare exception. Janet Monge as curator of the collection promotes its use in programs at the museum geared toward educating the general public about the history of American physical anthropology, as well as the history of race in science. Monge also allows members of the general public to engage the collection outside of formal programming (Monge, 2008; Renschler 2008; Renschler and Monge 2008).

---

[3]   Within the context of the critical and humanistic approach to science, the Descendant Community's involvement in the research process is a component of the democratization of knowledge that merges ethics and epistemology. Public vetting of the research design led to the use of particular language and research protocols that was valuable to the scope of the study. Therefore, public engagement made for a stronger research design with more interesting questions than those coming from researchers alone (Blakey, 2004).

This unique articulation of an anatomical and archaeological collection suggests a fate in which a greater degree of public engagement and politicization of the former is possible. What might that look like? Current mobilizations of data from both the NYABG and Cobb collections provide some examples and set a precedent for a fuller exploration of the contemporary relevance of anatomical collections.

For instance, the work of W. Montague Cobb and his skeletal collection are included in media and school-based public education initiatives. This includes his work being featured in documentaries such as Race: The Power of an Illusion (Pounder et al. 2003), as well as the author's school outreach efforts. The author and a colleague in cultural anthropology regularly visit elementary, middle and high schools to present the discipline of anthropology as a crucial part of understanding the world and effecting social change at local, national and global levels.[4] We draw upon our specialty areas to emphasize an understanding of what affects and maintains the biological and social health of communities. For instance, Cobb's work is used to illustrate biological anthropology's roots in anatomy and biology. We see this as an important part of creating a 'common sense logic' for students to understand the connection between anthropology and major educational initiatives such as those promoting science, technology, engineering and mathematics (STEM). STEM is at the forefront of a nationwide discourses promoting educational proficiency, scientific literacy, as well as those addressing economic uncertainty and social and economic mobility. To date, women hold only 25% of STEM jobs. According to the Digest of Education statistics, Only 4 percent of 9th graders eventually graduate with STEM degrees 10 years after entering high school. For African Americans students, the yield is between 1 and 2 percent (Chen 2013). These facts illustrate the difficulty that women and people of color have gaining access to STEM education and employment - often characterized as a 'leak' in the STEM pipeline.

Fatimah Jackson, Director of the Cobb Research Laboratory, recently announced a commitment to utilize the space to provide informal STEM learning opportunities to undergraduate students, graduate students and postdoctoral researchers (2015). The STEM related investigations will draw upon 20 generations of African American biological history from both the Cobb and NYABG collections. Data are consolidated into what Jackson and colleagues refer to as the 4Cs database, referring to the 400 years of African American biohistory encompassed in the span of time between the NYABG and Cobb collections (Jackson et al. 2016). The database is presented as ideal for interdisciplinary 'next generation' science research as a unique accumulation of historically contextualized Big Data 'on an underrepresented group known to have experienced differential survival over time' (Jackson et al. 2016, 510).

---

[4]   The general age categories for grades in the US educational system are as follows: elementary school is for children ages 5-10, Middle school is for children ages 11-13 and high school is for students ages 14-18. Outreach efforts are usually covered in local newspapers (for instance see https://www.insightnews.com/tag/kamela-heyward-rotimi/).

Conclusion: The Fate of Anatomical Collections in the US - Questions Worth Asking

Past and current mobilizations of the Cobb collection suggest that its fate will involve articulations that will preclude the structural violence that comes from not broadly exploring its relevance to current social, political, educational and scientific interests. If we are not limiting our examination of structural violence to the past treatment of people in life and death, then we are moved to question the notion of skeletal remains in the Cobb human archive and other collections existing in perpetuity as 'raw material' for the production of anthropological knowledge (TallBear 2011, 2013). This is turn begs the question of an uncritical regard for laboratory space as a 'natural' environment for remains to exist in perpetuity.

An examination of our current research practices in the context of structural violence also means that we must consider how to better integrate our research with broader intellectual and political discussions. Where do remains such as those in the Cobb collection fit into current discussions about the ethical treatment of African-American skeletal remains and artifacts? These discussions take place largely in the context of repatriation, protection and reburial, which seems to fall outside of the purview of anatomical collections. However, their always-already availability continues to place them at the center of anthropological knowledge production regarding health, disease and biological diversity. With this being the case, it follows that these collections should factor more heavily into these discussions. At the least, this would provide a means for a self-reflexive examination of the power relations between ourselves as researchers, the anatomical remains we study, and issues of accountability and community engagement. Returning to the New York African Burial Ground project, the Descendant Community factored heavily into operationalizing a critical and humanistic approach to studying the remains in terms of refining the research design and holding researchers accountable to public interests and concerns. More broadly, descendant communities are driving forces behind community activism associated with the repatriation, reburial and protection of African-American graves and cemeteries throughout the US (Dunnavant 2013; Mack 2007)[5]

How does structural violence, as well as principles of repatriation and public engagement allow for considering different spatial and relational possibilities for anatomical collections in the 21st century and beyond? If community collaboration is a basic standard for critical and humanistic research practices on non-anatomical remains, how might we establish mutually beneficial practices that include public engagement for future research on remains such as those in the Cobb collection? How

---

[5] The following websites were established by members of descendant communities and their allies to garner attention and support for the protection of African American cemeteries in Virginia and Washington, DC. This includes coordinating volunteer clean up days, political action and ritual commemoration of the burial sites. For instance, a ceremonial reading of the names of people buried at what is now Walter Pierce Park is held annually. https://walterpierceparkcemeteries.org/ and https://evergreencemeteryva.wordpress. com/. Efforts of the descendant community also garner attention in local and national newspapers, such as the one below focused on Evergreen cemetery in Richmond, VA: https://www.washingtonpost. com/local/virginia-politics/historic-black-cemeteries-seeking-the-same-support-virginia-gives-confederates/2017/02/11/888b8ec0-de67-11e6-acdf-14da832ae861_story.html?utm_term=.126ce5954a47.

does repatriation, protection and reburial directly bear upon collections of skeletal remains and documents like the Cobb human archive? What are the advantages, disadvantages and/or possibilities for research models that involve repatriating skeletal remains to identified family members? These are not new ideas or considerations. One of the stated aims of Blakey's curation of the Cobb collection was not only research, but burial or cremation establishing a protocol for identifying living descendants of individuals in the Cobb collection for possible repatriation (Blakey 2014). In addition, how might connections with descendants help to establish an ethical clientage to whom researchers are accountable – above and beyond our own scientific ethics (Blakey 1997; Blakey and Rankin-Hill 2004)?

In sum, the dual priorities of scientific rigor and social relevance should lead to conceptualizations of 'research progress' that view accountability and communication between scientists and the public as a moral obligation (Kakaliouras 2012). This dual obligation should also prioritize articulating the fates of the people in the collection with the fates of living people who are represented in these collections. Arguably, this will require multidisciplinary engagement that allows for both an historical and cultural reading of collections. Although our historical contextualizations of the remains we study are well developed, cultural readings continue to be lacking – especially those that consider the presence of African Americans in the collections. Arguably, we have yet to sufficiently address Michael Blakey's statement in his 2001 paper 'Bioarchaeology of the African Diaspora in the Americas: Its Origins and Scope' that: 'Bioanthropology is the cornerstone of African diasporic bioarchaeology, but the least informed by cultural and historical literatures' (Blakey 2001, 387). Making our research relevant to the living people whose histories are constructed by way of it requires articulation with intellectual work produced by and in the voices of those people. In the case developing cultural readings of US based anatomical collections, this includes the integration of scholarship from African American and African Diaspora studies to develop meaningful cultural analyses of collections that include African descendant remains. In terms of bioarchaeological studies, the New York African Burial Ground project discussed above remains a unique example of such an integration. And so, another aspect of mitigating structural violence in our current research practices involves the inclusion of cultural studies scholarship in the anthropological shaping of the past.

This brings the discussion full circle to my original point: studies of US based anatomical collections need to be open to multidisciplinary engagement that allows for identifying the power dynamics at work in shaping the past, and its relevance to the present. Structural violence has become a go-to framework for looking at these dynamics, but there continues to be an emphasis on looking at past structural violence that led to the particular composition of anatomical collections. However, we need to look more closely at how structural violence impacts current research and research practices. This includes its role in the continued dearth of people of color in the subfield. Drawing upon Galtung's and Farmer's language, there is a need for bringing ethnographic visibility to bioanthropological researchers in shaping the past, present and future of anatomical collections *as well as* the future of the field.

Acknowledgements

Thanks to the editors of this volume for their patience and vision, and kind regards to the other contributors to this volume. This paper is written in honor of past and present directors of the Cobb Laboratory. The paper is also written in honor of the students who began their bioanthropological training in the laboratory and now contribute to the field in a number of important ways.

References

Blakey, M. L. 1988. *The W. Montague Cobb Skeletal Collection: First Report.* Washington, D. C. Department of Sociology and Anthropology, Howard University.

Blakey, M. L. 1995. *The W. Montague Cobb Skeletal Collection and Biological Anthropology Laboratory at Howard University: Special Report and Announcement.* Washington, D. C., Department of Sociology and Anthropology, Howard University.

Blakey, M. L. 1997. Past is present: comments on 'In the realm of politics: prospects for public participation in African-American plantation archaeology'. *Historical Archaeology* 31(3): 140–145.

Blakey, M. L. 1998. Beyond European enlightenment: toward a critical and humanistic human biology. In A. H. Goodman and T. L. Leatherman (eds), *Building a New Biocultural Synthesis: Political-Economic Perspectives on Human Biology*: 389–405. Ann Arbor, University of Michigan Press.

Blakey, M. L. 2001. Bioarchaeology of the African diaspora: its origins and scope. *Annual Review of Anthropology* 30: 387–422.

Blakey, M. L. 2004. Theory: an ethical epistemology of publicly engaged biocultural research. In M. L. Blakey and L. Rankin-Hill (eds), *New York African Burial Ground: Skeletal Biology Report*: 17–28. Washington, D. C., Department of Sociology and Anthropology, Howard University; and Williamsburg, Virginia, The Institute for Historical Biology, Department of Anthropology, The College of William and Mary.

Blakey, M. L. 2010. African burial ground project: paradigm for cooperation? *Museum international* 62: 61–68.

Blakey, M. L., Leslie, T. E. and Reidy, J. P. 1994. Frequency and chronological distribution of dental enamel hypoplasia in enslaved African Americans: a test of the weaning hypothesis. *American Journal of Physical Anthropology* 95: 371–383.

Blakey, M. and Rankin-Hill, L. (eds) 2004. *New York African Burial Ground: Skeletal Biology Report.* Washington, D. C., Department of Sociology and Anthropology, Howard University; and Williamsburg, Virginia, The Institute for Historical Biology, Department of Anthropology, The College of William and Mary.

Chen, X. 2013. *STEM Attrition: College Students' Paths Into and Out of STEM Fields (NCES 2014-001).* Washington, D. C., National Center for Education Statistics, Institute of Education Sciences, U.S. Department of Education.

Cobb, W. M. 1934. The physical constitution of the American negro. *Journal of Negro Education* 3: 340–388.

Cobb, W. M. 1935. Municipal history from anatomical records. *Scientific Monthly* 40(2): 157–162.

Cobb, W. M. 1936. *The Laboratory of Anatomy and Physical Anthropology, Howard University, 1932-1936*. Washington, D. C., Howard University.

Cobb, W. M. n.d. *William Montague Cobb Manuscripts*. Washington, D. C., Moreland-Spingarn Research Center, Howard University.

de la Cova, C. 2010. Cultural patterns of trauma among 19th-century-born males in cadaver collections. *American Anthropologist* 112(4): 589–606.

de la Cova, C. 2011. Race, health, and disease in 19th-century-born males. *American Journal of Physical Anthropology* 144 (4): 526–537.

de la Cova, C. 2012. Trauma patterns in 19th-century-born African American and Euro-American females. *International Journal of Paleopathology* 2(2–3): 61–68.

Dunnavant, J. 2013. *Mortality Amongst Inhabited Alleys in the District of Columbia: A Case of the Mt. Pleasant Plains Cemetery*. Unpublished M.A. thesis, University of Florida.

Fabian, A. 2010. *The Skull Collectors: Race, Science, and America's Unburied Dead*. Chicago, University of Chicago Press.

Farmer, P. 2004. An anthropology of structural violence. *Current Anthropology* 45: 305–317.

Galtung, J. 1969. Violence, peace and peace research. *Journal of Peace Research* 6: 167–191.

Higgins, R., Haines, M., Walsh, L. and Sirianni, J. 2002. The poor in the mid 19th century Northeastern United States: evidence from the Monroe County Almshouse, Rochester, New York. In R. Steckel and J. Rose (eds), *Backbone of History: Health and Nutrition in the Western Hemisphere*: 162–183. New York, Cambridge University Press.

Jackson, F., Jackson, L., Cross, C. and Clarke, C. 2016. What could you do with 400 years of biological history on African Americans? Evaluating the potential scientific benefit of systematic studies of dental and skeletal materials on African Americans from the 17th through 20th centuries. *American Journal of Human Biology* 28: 510–513.

Jackson, F. 2015. Transforming the Cobb Research Laboratory into an Informal STEM Learning Resource Center. Available at : http://www.cobbresearchlab.com/blog/2015/12/18/transforming-the-cobb-research-laboratory-into-an-informal-stem-learning-resource-center.

Kakaliouras, A. M. 2012. An anthropology of repatriation: contemporary physical anthropological and Native American ontologies of practice. *Current Anthropology* 53: S210–S221.

Knoeff, R. and Zwijnenberg, R. 2015. Setting the stage. In R. Knoeff and R. Zwijnenberg (eds), *The Fate of Anatomical Collections*: 3–9. United Kingdom, Ashgate.

Knoeff, R. and Zwijnenberg, R. (eds) 2015. *The Fate of Anatomical Collections*. United Kingdom, Ashgate.

Lans, A. 2016. 'Whatever was once associated with him, continues to bear his stamp': articulating and dissecting George S. Huntington and his anatomical collection. Paper presented at the *Meetings of the American Anthropological Association*, Washington, D. C.

LaRoche, C. and Blakey, M. L. 1997. Seizing intellectual power: the dialogue at the New York African burial ground. *Historical Archaeology* 31: 84–106.

Mack, M. 2007. The public treatment of African American sacred space. Paper presented at the *Meetings of the American Anthropological Association*, Washington, D. C.

Monge, J. 2008. The Morton collection and NAGPRA. *Expedition Magazine* 50(3): 37.

Muller, J. 2006. *Trauma as a Biological Consequence of Inequality: A Biocultural Analysis of Washington D. C.'s African American Poor*. Unpublished PhD thesis, Department of Anthropology, University at Buffalo, Buffalo, New York.

Muller, J., Pearlstein, K. and de la Cova, C. 2016. Dissection and documented skeletal collections: embodiments of legalized inequality. In K. Nystrom (ed.), *The Bioarchaeology of Dissection and Autopsy in the United States*: 185–201. Switzerland, Springer.

Nystrom, K. 2014. Structural violence and dissection in the 19th-century United States. *American Anthropologist* 116: 765–779.

Nystrom, K. (ed.) 2016. *The Bioarchaeology of Dissection and Autopsy in the United States.* Switzerland, Springer.

Pearlstein, K. 2015. *Health and the Huddled Masses: An Analysis of Immigrant and Euro-American Skeletal Health in 19th Century New York City.* Unpublished PhD thesis, Department of Anthropology, American University.

Phillips, S. 2001. *Inmate Life in the Oneida County Asylum, 1860-1895: A Biocultural Study of the Skeletal and Documentary Evidence.* Unpublished PhD thesis, State University of New York at Albany.

Phillips, S. 2003. Worked to the bone: the biomechanical consequences of 'Labor Therapy' at a 19th-century asylum. In D. Herring and A. Swedlund (eds), *Human Biologists in the Archives*: 96–129. Cambridge, Cambridge University Press.

Pounder, C., Adelman, l., Cheng, J., Herbes-Sommers, C., Strain, T., Smith, l. and Ragazzi, C. 2003. *Race: the power of an illusion.* San Francisco, California Newsreel.

Rankin-Hill, L. and Blakey, M. 1994. W. Montague Cobb (1904–1990): physical anthropologist, anatomist, and activist. *American Anthropologist* 96: 74– 96.

Redman, S. 2016. *Bone Rooms: From Scientific Racism to Human Prehistory in Museums.* Cambridge, Harvard University Press.

Renschler, E. and Monge, J. 2008. The Samuel George Morton cranial collection: historical significance and new research. *Expedition Magazine* 50(3): 30–38.

Renschler, E. A. 2008. Historical osteobiography of the African crania in the Morton Collection. *Expedition Magazine* 50 (3): 33.

TallBear, K. (November 4, 2011). From Blood to DNA, From Tried to Race: Science, Whiteness and Property. [video file]. Retrieved from Cal Community Content, http://www.youtube.com/watch?v=UXScGwnX-1A.

TallBear, K. 2013. *Native American DNA: Tribal Belonging and the False Promise of Genetic Science.* Minneapolis, University of Minnesota Press.

Watkins, R. n.d. Anatomical collections as the bioanthropological other: some considerations. Paper presented at the *Meetings of the American Anthropological Association*, December 2, 2015, Denver, CO.

Watkins, R. 2003, To Know the Brethren: A Biocultural Profile of the W. Montague Cobb Skeletal Collection. Unpublished PhD thesis, Department of Anthropology, University of North Carolina, Chapel Hill, North Carolina.

Watkins, R. 2007. Knowledge from the margins: W. Montague Cobb's pioneering research in biocultural anthropology. *American Anthropologist* 109: 186–196.

Watkins R. 2010. Variation in health and socioeconomic status within the W. Montague Cobb skeletal collection: degenerative joint disease, trauma and cause of death. *International Journal of Osteoarchaeology* 22: 22–44.

Watkins, R. 2012. Biohistorical narratives of racial difference in the American Negro: notes toward a nuanced history of American physical anthropology. *Current Anthropology* 53: S196–S209.

Watkins, R. and Muller, J. 2015. Repositioning the Cobb human archive: the merger of a skeletal collection and its texts. *American Journal of Human Biology* 27: 41–50.

Zuckerman, M. K., Kamnikar, K. R. and Mathena, S. A. 2014. Recovering the 'body politic': a relational ethics of meaning for bioarchaeology. *Cambridge Archaeological Journal* 24: 513–522.

Chapter 10
Final Summary

Francisca Alves-Cardoso[1]

[1] LABOH – Laboratory of Biological Anthropology and Human Osteology, CRIA – Centro em Rede de Investigação em Antropologia, Faculdade de Ciências Sociais e Humanas, Universidade NOVA de Lisboa, Portugal.

This book originated from the idea of creating an interdisciplinary discussion on human identified skeletal collections focusing on their growing value, which has in turn led to a growing concern on all ethical issues surrounding their creation, and subsequent use in research and teaching. These latter aspects provided the title for the panel session "*Identified Skeletal Collections: The testing ground of anthropology?*" presented at the 17th World Congress of the International Union of Anthropological and Ethnological Sciences in Manchester, in August 2013, which was the starting point for the development of this book. At the time, the panel proposed to explore identified skeletal collections (ISC), understood as collections composed of skeletons, or parts of skeletons (i.e. bones) of named individuals, and their biographical data. The histories of the creation of ISC varies from the acquisition of the human remains and biographical data from recent cemeteries, via proper authorization sanctioned by the institutions responsible for cemetery management, donations and private collections, also including archaeologically excavated skeletons with name plates, as well as identified skeletons excavated in the context of war crimes, dictatorships, and politically repressive regimes. The majority of these scenarios were, to a certain extent, documented throughout the chapters in the book.

The argument supporting the creation and ongoing development of these collections is that they are privileged resources for developing methods to asses age-at-death and sex in skeletal remains, as well as allowing for the association of bone morphology to population ancestry, as well as the correlation of bone changes with diseases, and these are but a few of the singular areas of research that have been developed using remains and data from ISC. This is also a conforming idea presented and explored by all contributors of this book. The emphasis given to scientific knowledge provided by the collections is also strongly argued throughout these book contributions, and this is a view shared worldwide by all those that use human remains in teaching and research, or other associated activities. The high scientific value attributed to the collections exists regardless of the many biases inherent to the collections, which does not discourages their use. Alongside this, issues that may relate to ethical spects associated with the collection, amassing, curation and use of human remains and associated biographical data were implied. It was also interesting to verify that with the process of knowledge production being placed in various theoretical and academic backgrounds ranging from social and cultural sciences, alongside the recurrent biological and medical sciences, the skeletons that compose these collections find themselves being discussed as objects and subjects of science production, which adds another dimension to the overall conversation on ISC, and their constituents such as associated biographical data, forcing the need to bring an in-depth discussion on the matter to the forefront of science production.

Originally the panel had proposed the bringing of dialogue between various sciences and how these relate (in its broader sense) with ISC, including topics that were considered of future relevance to all aspects related with ISC. The topics stretched from the testing of ethnographical and osteological methods on ISC; exploring collection use and their limitations, with a focus on their history and how this history affected their usage in science and if it raised any specific social, curatorial or ethical issues. Societal aspects of the ISC, their importance for the comprehension of health, well-being and society in the past, any implications in the present and future were also mentioned. Importantly, the possibility of exploring ISC in an era of repatriation and reburial, in which ethical issues are of paramount importance was also possible. Overall, these were concerns that had been voiced in relation to other skeletal material of archaeological origin but had not found their way into identified collections, and/or identified remains incorporated in reference collections[1]. The knowledge of the personal and once living identity of the skeleton makes it more pressing to consider all ethical implication associated with the access to the remains, and associated data. This awareness of care, and preoccupation with identified collections has been changing in recent years, and consequently conversations on ethics and humans remains, either from identified collection or osteological collection driven from archaeological sites has been a recurrent theme. Nevertheless, and although a few years have passed since the discussion by the panel on *Identified Skeletal Collections: The testing ground of anthropology* (2013) many of the primary concerns continue to be worthy, and in need of in-depth discussion. For example, the debate on the ethics and the politics of identified human remains collected from modern cemeteries to build identified collections in Portugal was only recently addressed publicly, at a public debate entitled *Restless dead bodies: the ethics of circulation of human remains*, as part of the conference Bodies in Transition — Power, Knowledge and Medical Anthropology (EASA Medical Anthropology Network, 2017 Biennial Conference Network Meeting, that took place in Lisbon, Portugal in July 2017). The debate highlighted the news that the Lisbon municipality had given positive feedback on a donation of human skeletonized remains retrieved from the Lisbon municipal cemeteries to Simon Frazer University in Canada, to be used in scientific research and teaching. I consider this to be the starting point of a growing debate on the topic, which if properly dissected will indeed originate interesting questions worthy of prolonged and detailed debate.

The importance and significance of collections built based on human skeletal remains is acknowledged by all the authors that have contributed to this book. And, most importantly the many chapters exemplify the delicate stability that exists within science production, and the fact that collections associated with the assemblage of human remains used for teaching and research have become a

---

[1] Note that reference collections may not be composed of identified individuals. The concept of "reference collection" is used in association with collections that were/are used as a reference for the development of methods employed in the identification of demographical and other biological parameters in skeletons. Reference collections are referred to when addressing specific topics, and/or subjects they are associated with. Identified collections, which are also reference collection, are composed solely by identified skeleton, i.e. skeletons of which the biographical data is known.

major and valuable resource not only in universities, but also in association with museums and other public institutions. Furthermore, it becomes quite clear that the access to data is no longer restricted to the human remains, but extends to data associated with the remains including digital databases available online for use. Jelena Bekvalac's and Rebecca Redfern's chapter on the large human skeletal archaeological collections curated at the Museum of London, in the UK, exemplify these points to perfection. This chapter further highlights the role played by museums and associated research centers in public engagement as is the case of the Centre for Human Bioarchaeology at the Museum of London. The societal engagement with museum collections, amongst them the osteological collections, is informative as to their presence (the collections), their usefulness in research knowledge related with the evolutionary human history and past populations. Also, this public engagement allows for a sense of responsibility, i.e. the public not only becomes aware that these collections exist, and the manner in which human remains have been acquired and curated, but it also offers the opportunity of participation on society in this process and the possibility of the public to express their approval or disapproval on the matter.

Notwithstanding the value of the collections, it was interesting to observe that concerns exist as to their scientific viability. This is shared by almost all contributors. Many limitations and biases have been identified and explored alongside their problematic historical and socio-economic settings. On this specific topic, John Albanese's chapter entitled *Dealing with Bias in Identified Reference Collections and Implications for Research in the 21st Century* (Chapter 4) summarizes to perfection the major biases that may be found in, and associated with human collections, and how they may be addressed. The list is comprehensive, and placed in association with various collections globally making it clear that it is not an isolated issue. Albanese's concluding remark "Stripping individual skeletons and/or a collection as a whole of their context and reducing skeletons to ideal types existing outside of time and space, will necessarily lead to erroneous results, even if they are consistent with expectations in the discipline.". This is also reinforced in Chapter 8, exploring ethnographical approach to concepts used, and explored in association with skeletons from collections was able to demonstrate that biographical data at death is hardly representative of an individual life history. The lack of in-depth knowledge of an individual's life impacts on any interpretation made based on the observation of skeletal changes, and their associated with patterns of activity or health. On this specific aspect, it is interesting to consider that acknowledging the bias in the collections is a way forward to be able to explore these to the fullest, as well as all associated information, but and most importantly it is the perception that attempts made to correct these biases have introduced further biases. This is a vital consideration for anybody attempting to build a new collection or extend an existing one. Most importantly, by lifting the veil on the many biases and limitations associated with identified collections, it brings forward the question of the real contribution of these collections to science. Regardless, collections continue to be massively used in science production and in recent years with a strong association with forensic anthropology.

The book also demonstrates that ethical issues are a concern with all collections, but that they are addressed differently or are weighted differently across the globe, as it may be perceived in the many contributions in the book. It is always necessary to bare in mind that ethical treatment of human remains is a topic with a strong association with repatriation issues. This is not the context associated with the many identified skeletal collections or osteological collection which were built with remains from individuals not affected by repatriation issues, e.g. in Europe where the collections are amassed from locally living groups. Nowadays, ethical issues also consider   care, handling, access, research, display and retention of human skeletal remains in a location other than the place in which they were buried.

This overarching approach provides a new starting point and new avenues for discussion within the field of anthropology, and allied sciences. It highlights the need for a reflective approach to the collections, and discussion on the methodologies of grouping and use of human remains for research and training. It also places interest on the contribution that society may have in the construction of these collections emphasising the fact these are composed of identified individuals. With this in mind, human identified collections stand as a privileged testing ground for anthropological research and are an excellent interface between biology, society and culture.